Advances in Intelligent and Soft Computing 118

Editor-in-Chief: J. Kacprzyk

Advances in Intelligent and Soft Computing

Editor-in-Chief

Prof. Janusz Kacprzyk
Systems Research Institute
Polish Academy of Sciences
ul. Newelska 6
01-447 Warsaw
Poland
E-mail: kacprzyk@ibspan.waw.pl

Further volumes of this series can be found on our homepage: springer.com

Vol. 104. D. Jin and S. Lin (Eds.)
Advances in Computer Science, Intelligent System and Environment, 2011
ISBN 978-3-642-23776-8

Vol. 105. D. Jin and S. Lin (Eds.)
Advances in Computer Science, Intelligent System and Environment, 2011
ISBN 978-3-642-23755-3

Vol. 106. D. Jin and S. Lin (Eds.)
Advances in Computer Science, Intelligent System and Environment, 2011
ISBN 978-3-642-23752-2

Vol. 107. P. Melo-Pinto, P. Couto, C. Serôdio, J. Fodor, and B. De Baets (Eds.)
Eurofuse 2011, 2011
ISBN 978-3-642-24000-3

Vol. 108. Y. Wang (Ed.)
Education and Educational Technology, 2011
ISBN 978-3-642-24774-3

Vol. 109. Y. Wang (Ed.)
Education Management, Education Theory and Education Application, 2011
ISBN 978-3-642-24771-2

Vol. 110. L. Jiang (Ed.)
Proceedings of the 2011 International Conference on Informatics, Cybernetics, and Computer Engineering (ICCE 2011) November 19-20, 2011, Melbourne, Australia, 2011
ISBN 978-3-642-25184-9

Vol. 111. L. Jiang (Ed.)
Proceedings of the 2011 International Conference on Informatics, Cybernetics, and Computer Engineering (ICCE 2011) November 19-20, 2011, Melbourne, Australia, 2011
ISBN 978-3-642-25187-0

Vol. 112. L. Jiang (Ed.)
Proceedings of the 2011 International Conference on Informatics, Cybernetics, and Computer Engineering (ICCE 2011) November 19-20, 2011, Melbourne, Australia, 2011
ISBN 978-3-642-25193-1

Vol. 113. J. Altmann, U. Baumöl, and B.J. Krämer (Eds.)
Advances in Collective Intelligence 2011, 2011
ISBN 978-3-642-25320-1

Vol. 114. Y. Wu (Ed.)
Software Engineering and Knowledge Engineering: Theory and Practice, 2011
ISBN 978-3-642-03717-7

Vol. 115. Y. Wu (Ed.)
Software Engineering and Knowledge Engineering: Theory and Practice, 2011
ISBN 978-3-642-03717-7

Vol. 116. Yanwen Wu (Ed.)
Advanced Technology in Teaching - Proceedings of the 2009 3rd International Conference on Teaching and Computational Science (WTCS 2009), 2012
ISBN 978-3-642-11275-1

Vol. 117. Yanwen Wu (Ed.)
Advanced Technology in Teaching - Proceedings of the 2009 3rd International Conference on Teaching and Computational Science (WTCS 2009), 2012
ISBN 978-3-642-25436-9

Vol. 118. A. Kapczynski, E. Tkacz, and M. Rostanski (Eds.)
Internet - Technical Developments and Applications 2, 2012
ISBN 978-3-642-25354-6

Adrian Kapczynski, Ewaryst Tkacz,
and Maciej Rostanski (Eds.)

Internet - Technical Developments and Applications 2

Editors
Dr. Adrian Kapczynski
Silesian University of Technology
Akademicka Street 2a
44-100 Gliwice
Poland
E-mail: adrian.kapczynski@polsl.pl

Prof. Ewaryst Tkacz
Silesian University of Technology
Institute of Electronics
Division of Microelectronics
and Biotechnology
Akademicka Street 16
44-100 Gliwice
Poland
E-mail: etkacz@polsl.pl

Dr. Maciej Rostanski
Academy of Business in Dabrowa
Gornicza
Department of Computer Science
Cieplaka 1c
41-300 Dabrowa Gornicza
Poland
E-mail: mrostanski@wsb.edu.pl

ISSN 1867-5662 e-ISSN 1867-5670
ISBN 978-3-642-25354-6 e-ISBN 978-3-642-25355-3
DOI 10.1007/978-3-642-25355-3
Springer Heidelberg New York Dordrecht London

Library of Congress Control Number: 2011940870

Printed on acid-free paper

Springer is part of Springer Science+Business Media (www.springer.com)

Preface

In the period of last two decades it is not hard to observe unusual direct progress of civilization in many fields concerning conditionality coming up from technical theories or more generally technical sciences. We experience extraordinary dynamics of the development of technological processes including different fields of daily life which concerns particularly ways of communicating. We are aspiring for disseminating of the view that the success in the concrete action is a consequence of the wisdom won over, collected and appropriately processed. They are talking straight out about the coming into existence of the information society.

In such a context the meeting of the specialists dealing with the widely understood applications of the Internet give a new dimension associated with promoting something like the new quality. Because having the information in today's world of changing attitudes and socio-economic conditions can be perceived as one of the most important advantages. It results from the universal globalization letting observe oneself of surrounding world. Thanks to the development of the Internet comprehending the distance for broadcast information packages stopped existing. Also both borders of states don't exist and finally social, economic or cultural differences as well. It isn't possible to observe something like that with reference to no other means of communication. In spite of these only a few indicated virtues, how not arousing stipulations still it is much more to be done.

The monograph returned to hands of readers being a result of meeting specialists dealing with above mentioned issues should in the significant way contribute to the success in implementing consequences of human imagination into the social life. We believe being aware of a human weakness and an imperfection that the monograph presenting of a joint effort of the increasing numerically crowd of professionals and enthusiasts will influence the further technology development regarding IT with constantly expanding spectrum of its applications.

Dabrowa Gornicza,
November 2011

Adrian Kapczynski
Ewaryst Tkacz
Maciej Rostanski

Contents

Part I Mathematical and Technical Fundamentals

1 Local Controllability of Nonlinear Systems with Delays 3
Jerzy Klamka
1.1 Introduction 3
1.2 Preliminaries 4
1.3 Relative Controllability 7
1.4 Absolute Controllability 8
1.5 Semilinear Systems 9
1.6 Example 10
1.7 Conclusion 11
References 11

2 Comparison of Rules Synthesis Methods Accuracy in the System of Type 1 Diabetes Prediction 13
Rafal Deja
2.1 Introduction 13
2.2 Data Description 14
2.3 Problem Description 15
2.4 Methods 16
2.4.1 Rough Set Preliminaries 16
2.4.2 Rules Synthesis 17
2.4.3 Classification Accuracy 18
2.5 Results 19
2.5.1 RSES System 19
2.5.2 Being Sick Classifier 19
2.5.3 Diabetes Prediction Classifier 20
2.5.4 Diabetes Prediction Classification Accuracy 22
2.6 Conclusions 23
References 23

3 Progressive 3D Mesh Transmission in IP Networks: Implementation and Traffic Measurements 25
Slawomir Nowak and Damian Sobczak
3.1 Introduction 25
3.1.1 3D Objects' Transmission 26
3.2 Simulation Evaluation 27
3.2.1 Simulation Model and Simulation Results 27
3.2.2 Extension of the Simulation Model 29
3.3 Implementation 29
3.3.1 Preparation of Transmission Data 29
3.3.2 Client-Server Communication Protocol and Thresholds Description 31
3.4 3D Traffic Measurement 32
3.4.1 Experimental Results 32
3.5 Summary and Future Works 36
References 36

4 Model of the Network Physical Layer for Modern Wireless Systems 39
Maciej Wrobel and Slawomir Nowak
4.1 Introduction 39
4.2 The Model 41
4.3 Results 42
4.3.1 Propagation Model 43
4.3.2 Interference Model 44
4.3.3 Influence of the Mobility 46
4.4 Summary 47
References 48

5 Column-Oriented Metadata Organization of Vision Objects 49
Malgorzata Bach, Adam Duszenko, and Aleksandra Werner
5.1 Introduction 49
5.2 Problem Description 50
5.3 Solution Description 51
5.4 Tests 52
5.5 Summary 55
References 55

6 On Controllability of Linear Systems with Jumps in Parameters 57
Adam Czornik and Aleksander Nawrat
6.1 Introduction 57
6.2 Discrete-Time Jump Linear Systems 62
6.3 Controllability 66
6.4 Controllability with Respect to Expectation 67
6.5 Stochastic Controllability 69
6.5.1 Direct Controllability 71

6.6 Controllability at Random Time 75
6.7 Comparison and Discussion 79
References 80

Part II Information Management Systems and Project Management

7 General Description of the Theory of Enterprise Process Control 85
Miroslaw Zaborowski
7.1 The Purpose and the Scope of the EPC II Theory 85
7.2 Multilayer Structure of the Framework EPC II System 86
7.3 Business Processes and Functional Subsystems 87
7.4 Net of Transitions and Information Places of the Framework EPC II System 90
7.5 Relational Model of the Structure of Business Processes and Their Control Systems 90
7.6 Conclusions 92
References 92

8 On-Line Technological Roadmapping as a Tool to Implement Foresight Results in IT Enterprises 95
Andrzej M.J. Skulimowski and Przemyslaw Pukocz
8.1 Introduction 95
8.2 The Essence of Roadmapping 97
8.3 Formulating a Technological Strategic Planning Problem 99
8.4 Applying Roadmapping in Technological Investment Planning 102
8.5 An Application of Computer Assisted Roadmapping Support System to It Planning 105
8.6 Final Remarks 109
References 110

9 Software Testing in Systems of Large Scale 113
Wojciech Filipowski and Marcin Caban
9.1 The Role of Testing 113
9.2 Types or Classes of Tests 114
9.3 Test Planning and Choosing the Proper Test Set 116
9.4 Conclusions 119
References 119

10 Scope of Applications of the Framework Enterprise Process Control System 121
Miroslaw Zaborowski
10.1 The EPC II Theory 121
10.2 The Framework EPC II System as a Skeleton of the User's EPC II Systems 122

10.3 Cooperative EPC II Systems 125
10.4 The Framework EPC II System as a Simulator of Business Process Control 126
10.5 Conclusions 128
References 129

11 Selected Aspects of the Implementation Management in Multi-domain System Environment 131
Wojciech Filipowski and Marcin Caban
11.1 Organization of the Project Team 131
11.2 Planning - Problems with Synchronization of Plans among Different Domains versus the Schedule Baseline 132
11.3 Managing Unplanned Changes during the Implementation Process 134
11.4 Test Management 134
11.5 Quality Parameters (SLA) 137
11.6 Vendor Management (Contractual Issues - Costs, Penalties, Success Fee) 139
11.7 Conclusions 139
References 139

12 Electronic Recruitment System for University Applicants as B2C and B2P System 141
Paulina Puścian-Sobieska
12.1 Introduction 141
12.2 Computer Systems from Consumer Point If View 142
12.3 Classification of Electronic Recruitment System 143
12.4 Summary 146
References 146

Part III Information Security and Business Continuity Management

13 Computer Continuity Management in Distributed Systems of Management and Steering 149
Andrzej Grzywak and Piotr Pikiewicz
13.1 Introduction 149
13.2 Concept of a Measurement Method 151
13.3 Method of Measuring Data Analysis 152
13.4 Results of Conducted Measuring Research 152
13.5 Summary 154
References 154

14 Computer Support in Business Continuity and Information Security Management 155
Andrzej Bialas
14.1 Introduction 155
14.2 Basic Standards of the OSCAD Project 157

14.3 The Objectives of the OSCAD Project 158
14.4 The Range of Computer Support of Operations Related to the Management of Business Continuity and Information Security 159
14.5 General Concept of the OSCAD Software 159
14.6 Review of Modules Responsible for Business Continuity and Information Security Management 161
14.7 Computer Support in Business Continuity and Information Security Management 162
14.8 Review of Modules Responsible for Statistical Information Management and Other Auxiliary Modules 166
14.9 Conclusions 167
References 168

15 High Availability Methods for Routing in SOHO Networks 171
Maciej Rostanski
15.1 Introduction 171
15.2 Routing Redundancy with VRRP 172
15.2.1 Redundancy Solutions for Routing 173
15.2.2 VRRP Protocol 173
15.3 Simple Solution for Redundancy 175
15.3.1 Topology 175
15.3.2 Test Results 176
15.4 Conclusions 178
References 178

16 The Example of IT System with Fault Tolerance in a Small Business Organization 179
Pawel Buchwald
16.1 An Analysis of IT Systems Reliability In Order to Evaluate the Continuity of Business Services 179
16.2 Selected Mechanisms of Redundancy Used to Improve the Reliability 184
16.3 Conclusions 186
References 187

17 Local and Remote File Inclusion 189
Michal Hubczyk, Adam Domanski, and Joanna Domanska
17.1 Introduction 189
17.2 Local File Inclusion and Remote File Inclusion 191
17.2.1 Explanation of the Vulnerability 191
17.2.2 Example 1 191
17.2.3 Example 2 196
17.3 How to Protect 199
17.4 Conclusions 200
References 200

18 Maildiskfs - The Linux File System Based on the E-mails 201
Adam Domanski and Joanna Domanska
18.1 Introduction 201
18.2 The Linux Files System 202
18.3 The File System Maildiskfs 203
18.4 An E-mail Message 205
18.5 Performance Tests 206
18.6 Maildiskfs and the Spam Problem 207
18.7 Conclusions 207
References 208

19 Threats to Wireless Technologies and Mobile Devices and Company Network Safety 209
Teresa Mendyk-Krajewska, Zygmunt Mazur, and Hanna Mazur
19.1 Introduction 209
19.2 Threats to Wireless Technologies 210
19.3 Functionality of Placemobile Devices 212
19.4 Threats to Mobile Devices 214
19.5 Threats to Mobile Devices in Practice 217
19.6 Mobile Accessibility of Network Resources 221
19.7 Mobile Accessibility of Network Resources 222
19.8 Conclusion 224
References 224

20 Quantum E-Voting Cards 227
Marcin Sobota and Adrian Kapczynski
20.1 Introduction 227
20.2 Quantum Protocols 227
20.3 Protocol BB84 228
20.4 E-voting Protocols 230
20.5 Conception of Quantum E-voting Cards 231
20.6 Conclusion 232
References 232

21 Implementation of the OCTAVE Methodology in Security Risk Management Process for Business Resources 235
Marek Pyka and Ścibor Sobieski
21.1 Introduction 235
21.2 The OCTAVE Methodology 238
21.2.1 An Introduction to the OCTAVE Methodology 238
21.2.2 Characteristics of OCTAVE 239
21.2.3 Threats and Security Measures in OCTAVE 242
21.3 OCTAVE Workshops 243
21.3.1 Preparation 244
21.3.2 Phase 1: Build Asset-Based Threat Profiles 244

21.3.3 Phase 2: Identify Infrastructure Vulnerabilities 245
21.3.4 Phase 3: Develop Security Strategy and Plans 246
21.4 Example of OCTAVE Implementation 247
21.4.1 Systems and Databases 248
21.4.2 Personal Computers and Laptops 248
21.4.3 Documentation 249
21.4.4 Survey 249
21.4.5 Summary of Presented Example 250
21.5 Summary 251
References 251

22 Tools and Methods Used to Ensure Security of Processing and Storing Classified Information in Databases and IT Systems and Their Impact on System Performance 253
Łukasz Hoppe, Łukasz Wąsek, Arkadiusz Gwóźdź, and Aleksander Nawrat
22.1 Introduction 253
22.2 Classified Information – Legal Considerations 254
22.3 Ensuring Security of Classified Information – Tools and Methods 255
22.4 Physical Security 256
22.5 Software Protections 256
22.5.1 Consequences of Using Powerful Protections in Databases and IT Systems 258
22.6 Electronic Signature and Its Legal Aspect 261
22.7 Popularisation of PKI Technology in Poland 263
22.8 The Possibilities of Development of the Concept of PKI Application 265
References 267

Part IV Interdisciplinary Problems

23 Modern MEMS Acceleration Sensors in Tele-Monitoring Systems for Movement Parameters and Human Fall Remote Detection 271
Pawel Kostka and Ewaryst Tkacz
23.1 Introduction 271
23.2 Methods 272
23.2.1 MEMS Acceleration Sensors Principle of Operation 272
23.2.2 Mobile Data Acquisition System with Tele-Monitoring Modules 275
23.3 Results 276
23.4 Discussion and Conclusion 276
References 276

24 Tele-Manipulation System for Minimal Invasive Surgery Support. Prototype for Long Distance Operation 279
Pawel Kostka and Zbigniew Nawrat
24.1 Introduction 279
24.2 Methods 280
24.2.1 Robin Heart Vision (RHV)-Description of Surgery Manipulator and Mechanical Construction 280
24.2.2 Operation Field and Techniques Analysis for Robotic Supported Cardiac Surgery 281
24.2.3 Control System for Surgery Telemanipulator: Local Tele-Manipulation 281
24.2.4 Long Distance Tele-Manipulation Experiment 283
24.3 Results 285
24.4 Discussion and Conclusion 286
References 286

List of Contributors

Malgorzata Bach
Faculty of Automatic Control, Electronics and Computer Science, Silesian University of Technology, Akademicka 16, 44-100 Gliwice, Poland
e-mail: malgorzata.bach@polsl.pl

Andrzej Bialas
Research and Development Centre EMAG, Leopolda 31, 40-189 Katowice, Poland
e-mail: a.bialas@emag.pl

Pawel Buchwald
Department of Computer Science, Academy of Business in Dabrowa Gornicza, Cieplaka 1c, 41-300 Dabrowa Gornicza, Poland
e-mail: pawel@buchwald.com.pl

Marcin Caban
Faculty of Automatic Control, Electronics and Computer Science, Silesian University of Technology, Akademicka 16, 44-100 Gliwice, Poland
e-mail: marcin.caban@polsl.pl

Adam Czornik
Silesian University of Technology, Gliwice, Poland
e-mail: Adam.Czornik@polsl.pl

Adam Duszenko
Faculty of Automatic Control, Electronics and Computer Science, Silesian University of Technology, Akademicka 16, 44-100 Gliwice, Poland
e-mail: adam.duszenko@polsl.pl

Rafal Deja
Department of Computer Science, Academy of Business in Dabrowa Gornicza, Cieplaka 1c, 41-300 Dabrowa Gornicza, Poland
e-mail: rdeja@wsb.edu.pl

Joanna Domanska
Institute of Theoretical and Applied Informatics, Polish Academy of Sciences, Baltycka 5, 44-100 Gliwice, Poland
e-mail: joanna@iitis.gliwice.pl

Department of Computer Science, Academy of Business in Dabrowa Gornicza, Cieplaka 1c, 41-300 Dabrowa Gornicza, Poland
e-mail: jdomanska@wsb.edu.pl

Adam Domanski
Department of Computer Science, Academy of Business in Dabrowa Gornicza, Cieplaka 1c, 41-300 Dabrowa Gornicza, Poland
e-mail: adam.domanski@polsl.pl

Wojciech Filipowski
Faculty of Automatic Control, Electronics and Computer Science, Silesian University of Technology, Akademicka 16, 44-100 Gliwice, Poland
e-mail: wojciech.filipowski@polsl.pl

Andrzej Grzywak
Department of Computer Science, Academy of Business in Dabrowa Gornicza, Cieplaka 1c, 41-300 Dabrowa Gornicza, Poland
e-mail: agrzywak@wsb.edu.pl

Arkadiusz Gwóźdź
WASKO S. A., Berbeckiego 6, 44 - 100 Gliwice, Poland

Łukasz Hoppe
WASKO S. A., Berbeckiego 6, 44 - 100 Gliwice, Poland

Michal Hubczyk
Institute of Theoretical and Applied Informatics, Polish Academy of Sciences, Baltycka 5, 44-100 Gliwice, Poland,

Adrian Kapczynski
Department of Computer Science, Academy of Business in Dabrowa Gornicza, Cieplaka 1c, 41-300 Dabrowa Gornicza, Poland
e-mail: akapczynski@wsb.edu.pl

Jerzy Klamka
Department of Computer Science, Academy of Business in Dabrowa Gornicza, Cieplaka 1c, 41-300 Dabrowa Gornicza, Poland
e-mail: jerzy.klamka@polsl.pl

Pawel Kostka
Institute of Electronics, Faculty of Automatic Control, Electronics and Computer Science, Silesian University of Technology, Akademicka 16, 44-100 Gliwice, Poland
e-mail: pawel.kostka@polsl.pl

Hanna Mazur
Institute of Informatics, Wroclaw University of Technology, Wybrzeze Wyspianskiego 27, 50-370 Wroclaw, Poland
e-mail: hanna.mazur@pwr.wroc.pl

Zygmunt Mazur
Institute of Informatics, Wroclaw University of Technology, Wybrzeze Wyspianskiego 27, 50-370 Wroclaw, Poland
e-mail: zygmunt.mazur@pwr.wroc.pl

Teresa Mendyk-Krajewska
Institute of Informatics, Wroclaw University of Technology, Wybrzeze Wyspianskiego 27, 50-370 Wroclaw, Poland
e-mail: teresa.mendyk-krajewska@pwr.wroc.pl

Aleksander Nawrat
WASKO S. A., Berbeckiego 6, 44 - 100 Gliwice, Poland
e-mail: Aleksander.Nawrat@polsl.pl

Zbigniew Nawrat
Institute of Heart Prostheses, Foundation for Cardiac Surgery Development, Zabrze, Poland
e-mail: nawrat@frk.pl

Slawomir Nowak
Department of Computer Science, Academy of Business in Dabrowa Gornicza, Cieplaka 1c, 41-300 Dabrowa Gornicza, Poland
e-mail: snowak@wsb.edu.pl

Piotr Pikiewicz
Department of Computer Science, Academy of Business in Dabrowa Gornicza, Cieplaka 1c, 41-300 Dabrowa Gornicza, Poland
e-mail: ppikiewicz@wsb.edu.pl

Przemyslaw Pukocz
AGH University of Science and Technology, Chair of Automatic Control, Decision Science Laboratory, Krakow, Poland
e-mail: pukocz@agh.edu.pl

Paulina Puscian-Sobieska
Informatic Center, University of Łodź, Lindleya 3 st., 90-131 Łódź, Poland
e-mail: paulina@uni.lodz.pl

Marek Pyka
Department of Computer Science, Academy of Business in Dabrowa Gornicza, Cieplaka 1c, 41-300 Dabrowa Gornicza, Poland
e-mail: mpyka@wsb.edu.pl

Maciej Rostanski
Department of Computer Science, Academy of Business in Dabrowa Gornicza, Cieplaka 1c, 41-300 Dabrowa Gornicza, Poland
e-mail: mrostanski@wsb.edu.pl

Andrzej M.J. Skulimowski
AGH University of Science and Technology, Chair of Automatic Control, Decision Science Laboratory, Krakow, Poland
e-mail: ams@agh.edu.pl

Damian Sobczak
Institute of Theoretical and Applied Informatics, Polish Academy of Sciences, Baltycka 5, 44-100 Gliwice, Poland,

Scibor Sobieski
Department of Theoretical Physics and Informatics, Faculty of Physics and Applied Informatics, University of Lodz, ul. Pomorska 149/153, 90-236 Lodz, Poland
e-mail: scibor@uni.lodz.pl

Marcin Sobota
Department of Computer Science, Academy of Business in Dabrowa Gornicza, Cieplaka 1c, 41-300 Dabrowa Gornicza, Poland
e-mail: ppikiewicz@wsb.edu.pl

Ewaryst Tkacz
Institute of Electronics, Faculty of Automatic Control, Electronics and Computer Science, Silesian University of Technology, Akademicka 16, 44-100 Gliwice, Poland
e-mail: etkacz@polsl.pl

Aleksandra Werner
Faculty of Automatic Control, Electronics and Computer Science, Silesian University of Technology, Akademicka 16, 44-100 Gliwice, Poland
e-mail: aleksandra.werner@polsl.pl

Łukasz Wąsek
WASKO S. A., Berbeckiego 6, 44 - 100 Gliwice, Poland

Maciej Wrobel
Department of Computer Science, Academy of Business in Dabrowa Gornicza, Cieplaka 1c, 41-300 Dabrowa Gornicza, Poland
e-mail: wrobelmaciek@gmail.com

Miroslaw Zaborowski
Department of Computer Science, Academy of Business in Dabrowa Gornicza, Cieplaka 1c, 41-300 Dabrowa Gornicza, Poland
e-mail: mzaborowski@wsb.edu.pl

Part I
Mathematical and Technical Fundamentals

Chapter 1
Local Controllability of Nonlinear Systems with Delays

Jerzy Klamka

Abstract. In the present paper local constrained controllability problems for nonlinear system with constant delays in the control are formulated and discussed. Using some mapping theorems taken from functional analysis and linear approximation methods sufficient conditions for relative and absolute local constrained controllability in a given time interval are derived and proved. The present paper extends controllability conditions with unconstrained controls given in the literature to cover the case of nonlinear systems with delays in control and with constrained controls. Special case of semilinear systems without delays is also considered.

1.1 Introduction

Controllability is one of the fundamental concept in mathematical control theory. Systematic study of controllability was started at the beginning of sixties, when the theory of controllability based on the description in the form of state space for both time-invariant and time-varying linear control systems was worked out. Roughly speaking, controllability generally means, that it is possible to steer dynamical control system from an arbitrary initial state to an arbitrary final state in certain time interval using the set of admissible controls [8]. In recent years various controllability problems for different types of linear dynamical systems have been considered in many publications and monographs. The extensive list of these publications can be found for example in the monograph [8]. However, it should be stressed, that the most literature in this direction has been mainly concerned with deterministic controllability problems for finite-dimensional linear dynamical systems with unconstrained controls and without delays.

Jerzy Klamka
Academy of Business in Dabrowa Gornicza, Department of Computer Science
ul. Cieplaka 1c, 41-300 Dabrowa Gornicza, Poland
e-mail: jerzy.klamka@polsl.pl
http://www.wsb.edu.pl

A. Kapczynski et al. (Eds.): Internet - Technical Develop. & Appli. 2, AISC 118, pp. 3–12.
springerlink.com

Up to the present time the problem of controllability in continuous and discrete time linear dynamical systems has been extensively investigated in many papers (see e.g., [1], [3], or [5] for the extensive list of publications).

However, this is not true for the nonlinear dynamical systems specially with delays in control and with constrained controls. Only a few papers concern constrained controllability problems for continuous or discrete nonlinear dynamical systems.

In the present paper local relative and absolute constrained controllability problems for nonlinear stationary system with many constant delays in control are formulated and discussed.

Using some mapping theorems taken from functional analysis [2], [5] and linear approximation methods sufficient conditions for constrained relative and absolute controllability are derived and proved. Local controllability for special case of semilinear stationary dynamical systems without delays in control is also considered.

The present paper extends in some sense the results given in papers [2], [4] and [5] to cover the nonlinear systems with delays in control and with constrained controls.

1.2 Preliminaries

Let us consider general nonlinear system with constant delays described by the following differential equation

$$x'(t) = f(x(t), u(t-h_0), u(t-h_1), ..., u(t-h_i), ..., u(t-h_M)) \qquad (1.1)$$

where: $0 = h_0 < h_1 < ... < h_i < ... < h_M$ are constant delays,

$x(t) \in R^n$ is a state vector the time t,

$u(t) \in R^m$ is a control vector at the time t,

$f : R^n \times R^m \times R^m \times ... \times R^m \times ... \times R^m \to R^n$, is a given function, which is continuously differentiable with respect all its arguments in some neighborhood of zero and

$$f(0,0,...,0,...,0) = 0.$$

Let $U \subset R^m$ be a given arbitrary set. Let $L_\infty([0,t_1],U)$ be the set of admissible controls. In the sequel we shall also use the following notations: $\Omega^0 \subset R^m$ is a neighborhood of zero, $U^c \subset R^m$ is a closed convex cone with vertex at zero and $U^{c0} = U^c \frown \Omega^0$. Moreover, let us observe, that for $u = U^c$ the set of admissible controls $L_\infty([0,t_1],U^c)$ is a cone in the linear space $L_\infty([0,t_1],R^m)$.

The initial conditions for nonlinear differential equation 1.1 are given by

$$x(0) = x_0 \in R^n \quad u_0 \in L_\infty([-h_M,0],U) \qquad (1.2)$$

where x_0 is a known vector and u_0 is a known function.

For a given initial conditions 1.2 and for an arbitrary admissible control $u_0 \in L_\infty([0,t_1],U)$ there exists unique solution of the nonlinear differential equation 1.1 $x(t;x_0,u_0,u) \in R^n$, for $t \in [0,t_1]$.

Since function f is continuously differentiable therefore, using the standard methods, it is possible to construct linear approximation of the nonlinear system 1.1. This linear approximation is valid in some neighborhood of the point zero in the product space $R^n \times R^m \times R^m \times ... \times R^m \times ... \times R^m$, and is given by the linear differential equation 1.3

$$x'(t) = Ax(t) + \sum_{i=o}^{i=M} B_i u(t-h_i) \tag{1.3}$$

defined for $t \geq 0$, where A is $n \times n$-dimensional constant matrix, and $B_i, i = 0,1,2,...,M$ are $n \times m$-dimensional constant matrices given by:

$$A = f'_x(0,0,0,...,0,...,0) \quad ,$$

$$B_i = f'_{u_i}(0,0,0,...,0,...,0), \quad for \quad i = 0,1,2,...,M$$

For linear system 1.3 we can define the so called state transition matrix exp(At). Using the state transition matrix exp(At) we can express the solution $x(t;x_0,u_0,u)$ of linear system (2.3) for $t > h_M$ in the following compact form

$$x(t;x_0,u_0,u) = \exp(At)x_0 +$$
$$+ \int_0^t \exp(A(t-\tau)) \left(\sum_{i=o}^{i=M} B_i u(\tau - h_i) \right) d\tau =$$
$$= \exp(At)x_0 + \sum_{i=o}^{i=M} \int_{-h_i}^{0} \exp(A(t+s+h_i))B_i u_0(s)ds +$$
$$+ \sum_{i=o}^{i=M} \int_0^t \exp(A(t+s+h_i))B_i u(s)ds \tag{1.4}$$

For zero initial condition

$$x(0) = x_0 = 0, \quad u_0 = 0,$$

the solution $x(t;0,0,u)$ to 1.3 for $t > h_M$ is given by:

$$x(t;0,0,u) = \sum_{i=o}^{i=M} \int_0^t \exp(A(t+s+h_i))B_i u(s)ds \tag{1.5}$$

Finally, let us also define the associated linear system without delays

$$x'(t) = Ax(t) + Du(t) \tag{1.6}$$

where $D = \sum_{i=o}^{i=M} \exp(Ah_i)B_i$

For linear and nonlinear systems with delays in control it is possible to define many different concepts of controllability [2], [3], [4]. In the sequel, we shall concentrate on local and global relative and absolute U-controllability in a given time interval $[0,t_1]$, where $t_1 > h_M$.

Definition 2.1. System (1) is said to be globally relatively U-controllable in a given interval $[0,t_1]$, if for zero initial conditions $x0 = 0, u0 = 0$ and every vector $x' \in R^n$, there exists an admissible control $u \in L_\infty([0,t_1],U)$, such that the corresponding solution of the equation (1) satisfies condition $x(t_1,0,0,u) = x'$.

Definition 2.2. System (1) is said to be locally relatively U-controllable in a given interval $[0,t_1]$, if for zero initial conditions $x0 = 0, u0 = 0$, there exists neighborhood of zero $D \subset R^n$, such that for every point $x' \in D$ there exists an admissible control $u \in L_\infty([0,t_1],U)$, such that the corresponding solution of the equation (1) satisfies the condition $x(t_1,0,0,u) = x'$.

Definition 2.3. System (1) is said to be globally absolutely U-controllable in a given interval $[0,t_1]$, if for zero initial conditions $x0 = 0, u0 = 0$, any given function $u_1 \in L_\infty([t_1 - h_M,t_1],U)$, and every vector $x' \in R^n$, there exists an admissible control $u \in L_\infty([0,t_1 - h_M],U)$, such that the corresponding solution of the equation (1) satisfies condition $x(t_1,0,0,u) = x'$.

Definition 2.4. System (1) is said to be locally absolutely U-controllable in a given interval $[0,t_1]$ if for zero initial conditions $x0 = 0, u0 = 0$, and any given function $u_1 \in L_1([t_1 - h_M,t_1],U)$ there exists neighborhood of zero $D \subset R^n$, such that for every point $x' \in D$ there exists an admissible control $u \in L_\infty([0,t_1 - h_M],U)$, such that the corresponding solution of the equation (2.1) satisfies the condition $x(t_1,0,0,u) = x'$.

Of course the same definitions are valid for linear systems (3). For linear systems with delays in control various controllability conditions are presented in the literature (see e.g., [1], [3] or [4]). It is well known [2], that for the sets U containing zero as an interior point, the local relative (absolute) constrained controllability is equivalent to the global relative (absolute) unconstrained controllability.

Lemma 2.1. [2] Linear system (3) is locally relative (absolute) ?0-controllable in the time interval $[0,t_1]$ if and only if it is globally relative (absolute) R^m-controllable in the time interval $[0,t_1]$.

Corollary 2.1 directly follows from the well known fact, that the range of the linear bounded operator covers whole space if and only if this operator transforms some neighborhood of zero onto some neighborhood of zero in the range space [1].

Finally, it should be pointed out, that absolute local (global) constrained controllability of the linear system with delays (3) in the time interval $[0,t_1]$ is equivalent to local (global) constrained controllability in the time interval $[0,t_1 - h_M]$ of the associated linear system without delays (6).

Lemma 2.2. [4] Linear system (3) is locally (globally) absolute U-controllable in the interval $[0,t_1]$ if and only if the associated linear system without delays is locally (globally) U-controllable in the interval $[0,t_1-h_M]$.

1.3 Relative Controllability

In this section we shall formulate and prove sufficient conditions of the local U-controllability in a given interval $[0,t_1]$ with different sets U for the nonlinear system (1). Proofs of the main results are based on some lemmas taken directly from functional analysis and concerning so called nonlinear covering operators [1], [5]. Now, for convenience we shall shortly state this results.

Lemma 2.1. [2], [5] Let $F: Z \rightarrow Y$ be a nonlinear operator from a Banach space Z into a Banach space Y and suppose that $F(0)=0$. Assume that the Frechet derivative $dF(0)$ maps a closed convex cone $C \subset Z$ with vertex at zero onto the whole space Y. Then there exists neighborhoods $M_0 \subset Z$ about $0 \in Z$ and $N_0 \subset Y$ about $0 \in Y$ such that the equation $y=F(z)$ has for each $y \in N_0$ at least one solution $z \in M_0 \frown C$.

Let us observe, that a direct consequence of Lemma 3.1 is the following result concerning nonlinear covering operators.

Lemma 2.2. [2], [5] Let $F: Z \rightarrow Y$ be a nonlinear operator from a Banach space Z into a Banach space Y which has the Frechet derivative $dF(0): Z \rightarrow Y$, whose image coincides with the whole space Y. Then the image of the operator F will contain a neighborhood of the point $F(0) \in Y$.

Now, we are in the position to formulate and prove the main result on the local relative U-controllability in the interval $[0,t_1]$ for the nonlinear system (1).

Theorem 2.1. Let us suppose, that $U^c \subset R^m$ is a closed convex cone with vertex at zero. Then the nonlinear system (1) is locally relatively U^{c0}-controllable in the interval [0,t1] if its linear approximation near the origin given by the differential equation (3) is globally relatively Uc-controllable in the same interval $[0,t_1]$.

Proof. Proof of the Theorem 2.1 is based on Lemma 2.1. Let our nonlinear operator F transforms the space of admissible controls $L_\infty([0,t1],Uc)$ into the space Rn at the time t_1 for the nonlinear system (1). More precisely, the nonlinear operator $F:$ $R^m \times R^m \times R^m \times \ldots \times R^m \rightarrow R^n$ is defined as follows

$$F(L_\infty([0,t_1],U^c)) = x(t_1,0,0,u) \tag{1.7}$$

where $x(t_1,0,0,u)$ is the solution at time $t_1 > h_M$ of the nonlinear system (1) corresponding to an admissible control $u \in L_\infty([0,t_1],U^c)$ and for zero initial conditions. Frechet derivative at point zero of the nonlinear operator F denoted as dF(0) is a linear bounded operator defined by the following formula

$$dF(0)(L_\infty([0,t_1],U^c)) = x(t_1,0,0,u) \tag{1.8}$$

where $x(t_1,0,0,u)$ is the solution at time t_1 of the linear system (3) corresponding to an admissible control $u \in L_\infty([0,t_1],U^c)$ and for zero initial conditions.

Since f(0,0,...,0,...,0)=0, then for zero initial conditions nonlinear operator F transforms zero into zero i.e., F(0)=0. If linear system (3) is globally relatively U^c-controllable in the interval $[0,t_1]$, then the image of Frechet derivative dF(0) covers whole space R^n. Therefore, by the result stated at the beginning of the proof, the nonlinear operator F covers some neighborhood of zero in the space R^n. Hence, by Definition 2.2 nonlinear system (1) is locally relatively U^c-controllable in the interval $[0,t_1]$.

In the case when the set U contains zero as an interior point, then by Lemma 3.2 we have the following sufficient condition for local constrained relative controllability of nonlinear system (1).

Corollary 2.1. Let $0 \in int(U)$ .Then the nonlinear system (3.1) is locally relatively U-controllable in the interval $[0,t_1]$ if its linear approximation near the origin given by the differential equation (3) is globally relatively Rm-controllable in the same interval $[0,t_1]$.

1.4 Absolute Controllability

Using similar methods as in Section 3 it is possible to derive sufficient conditions for local absolute U^c-controllability in a given interval $[0,t_1], t_1 > h_M$ for the nonlinear system (2.1).

Theorem 3.1. Let us suppose, that $U^c \subset R^m$ is a closed convex cone with vertex at zero. Then the nonlinear system (1) is locally absolutely U^{c0}-controllable in the interval $[0,t_1]$ if its linear approximation near the origin given by the differential equation (3) is globally absolutely U^{c0}-controllable in the same interval $[0,t_1]$.

Proof. Proof of the Theorem 3.1 is similar to the proof of Theorem 3.1 and is based on Lemma 2.1. Let our nonlinear operator F transforms the space of admissible controls $L_\infty([0,t_1-h_M],U^c)$ into the space R^n at the time t_1-h_M for the nonlinear system (1). More precisely, the nonlinear operator $F: R^m \times R^m \times ... \times R^m \times R^n$ is defined as follows

$$F(L_\infty([0,t_1],U^c)) = x(t_1-h_M,0,0,u) \tag{1.9}$$

where $x(t_1-h_M,0,0,u)$ is the solution at time t_1-h_M of the nonlinear system (1) corresponding to an admissible control $u \in L_\infty([0,t_1-h_M],U^c)$ and for zero initial conditions. Frechet derivative at point zero of the nonlinear operator F denoted as dF(0) is a linear bounded operator defined by the following formula

$$dF(0)(L_\infty([0,t_1-h_M],U^c)) = x(t_1-h_M,0,0,u) \tag{1.10}$$

where $x(t_1-h_M,0,0,u)$ is the solution at time t_1 of the linear system (3) corresponding to an admissible control $u \in L_\infty([0,t_1-h_M],U^c)$ and for zero initial conditions.

Since f(0,0,...,0,...,0)=0, then for zero initial conditions nonlinear operator F transforms zero into zero i.e., F(0)=0. If linear system (3) is globally absolutely U^c-controllable in the interval $[0,t_1]$, then the image of Frechet derivative dF(0) covers whole space R^n. Therefore, by the result stated at the beginning of the proof, the nonlinear operator F covers some neighborhood of zero in the space R^n. Hence, by Definition 2.2 nonlinear system (1) is locally absolutely U^c-controllable in the interval $[0,t_1]$.

In the case when the set U contains zero as an interior point, then by Lemma 3.2 we have the following sufficient condition for local constrained absolute controllability of nonlinear system (7).

Corollary 4.1. Let $0 \in int(U)$.Then the nonlinear system (7) is locally absolutely U-controllable in the interval $[0,t_1]$ if its linear approximation near the origin given by the differential equation (3) is globally absolutely R^m-controllable in the same interval $[0,t_1]$.

1.5 Semilinear Systems

Semilinear dynamical systems are special and important cases of general nonlinear dynamical systems. In this section using the results of the previous sections, we study special case of general nonlinear system (1) with delays in controls namely, the semilinear stationary finite-dimensional control system without delays in control described by the following ordinary differential state equation

$$x'(t) = Ax(t) + F(x(t),u(t)) + Bu(t) \quad for \quad t \in [0,t_1] \tag{1.11}$$

with zero initial conditions: x(0) = 0

where the state $x(t) \in R^n$
the control $u(t) \in R_m$,
A is $n \times n$ dimensional constant matrix,
B is $n \times m$ dimensional constant matrix

Moreover, let us assume that the nonlinear mapping

$$F : R^n \times R^m \to R^n$$

is continuously differentiable near the origin and such that F(0,0)=0.

$$z'(t) = Cz(t) + Du(t) \quad for \quad t \in [0,t_1] \tag{1.12}$$

with zero initial condition z(0)=0, where

$$C = A + D_F(0,0) \quad D = B + D_uF(0,0) \tag{1.13}$$

are $n \times n$-dimensional and $n \times m$-dimensional constant matrices, respectively.

The main result of this section is the following sufficient condition for constrained local controllability of the semilinear dynamical system (11).

Theorem 5.1 Suppose that

(i) $F(0,0) = 0$,
(ii) $U_c \subset R^m$ is a closed and convex cone with vertex at zero,
(iii) The associated linear control system (31) is U_c-globally controllable in $[0,t_1]$.

Then the semilinear stationary dynamical control system (1) is U_c-locally controllable in $[0,t_1]$.

It should be pointed out, that in applications of the Theorem 5.1, the most difficult problem is to verify the assumption (iii) about constrained global controllability of the linear stationary dynamical (12). In order to avoid this disadvantage, we may use the following well known Theorem 5.2.

Theorem 5.2 [1], [2]. Suppose the set U_c is a cone with vertex at zero and nonempty interior in the space R^m.

Then the associated linear stationary dynamical control system (12) is U_c-globally controllable in time interval $[0,t_1]$ if and only if

(1) it is controllable without any constraints, i.e.
$rank[D, CD, C^2D, ..., C^{n-1}D] = n$,
(2) there is no real eigenvector $v \in R^n$ of the matrix C^{tr} satisfying inequalities
$v^{tr}Du \leq 0, for \quad all \quad u \in U_c$.

It should be pointed out that for the special case namely, for the single input scalar admissible control local constrained controllability conditions for linear system (12) are rather simple. In this case i.e., for the case m=1, Theorem 5.2 reduces to the following Corollary.

Corollary 5.1. [1], [1]. Suppose that $m = 1$ and $U_c = R^+$.

Then the associated linear dynamical control system (31) is U_c-globally controllable in $[0,t_1]$ if and only if it is controllable without any constraints i.e.,
$rank[D, CD, C^2D, ..., C^{n-1}D] = n$,
and matrix C has only complex eigenvalues.

1.6 Example

In this section as an illustrative example we shall consider constrained local controllability of models C, described by the following semilinear differential state equations:

$$x' = y - x,$$
$$y' = -\vartheta(e^{2/3x} - 1) + \rho u$$

Therefore, taken into account the general form of seimilinear dynamic systems we have

$$A = \begin{bmatrix} -1 & 1 \\ 0 & 0 \end{bmatrix}$$

$$F(x,u) = F(x) = \begin{bmatrix} 0 \\ -\vartheta(e^{2/3x} - 1) \end{bmatrix}$$

$$B = \begin{bmatrix} 0 \\ \rho \end{bmatrix}$$

Hence, we have

$$F_x(0,0) = \begin{bmatrix} 0 & 0 \\ \vartheta 2/3 & 0 \end{bmatrix}$$

$$C = A + F_x(0,0) = \begin{bmatrix} -1 & 1 \\ \vartheta 2/3 & 0 \end{bmatrix}$$

$$rank[B \quad CB] = rank = \begin{bmatrix} 0 & \rho \\ \rho & 0 \end{bmatrix} = 2 = n$$

Moreover, the characteristic polynomial for the matrix C has the following form:

$$det(sI - C) = det \begin{bmatrix} s+1 & -1 \\ \vartheta 2/3 & s \end{bmatrix} = s(s+1) + 2/3\vartheta = s^2 + s + 2/3\vartheta$$

Hence, since $\Delta = 1 - 8/3\vartheta$, then for $\vartheta > 3/8$ and matrix C has only complex eigenvalues, and considered dynamical system without delays is constrained locally controllable in any time interval..

1.7 Conclusion

In the present paper different types of controllability of nonlinear control system with constant delays in control has been considered. The results presented can be extended in many directions. For example it is possible to formulate sufficient local controllability conditions for nonlinear time-varying systems. Moreover, similar controllability results can be derived for very general nonlinear systems with distributed and several time-dependent delays in the control and for nonlinear infinite-dimensional systems with different kinds of delays.

Acknowledgements. This work was supported by National Science Foundation under grant no. 4423/B/T02/2010/39.

References

1. Klamka, J.: CControllability of Dynamical Systems. Kluwer Academic Publishers, Dordrecht (1991)
2. Klamka, J.: Constrained controllability of nonlinear systems. Journal of Mathematical Analysis and Applications 201(2), 365–374 (1996)

3. Klamka, J.: Stochastic controllability of systems with multiple delays in control. International Journal of Applied Mathematics and Computer Science 19(1), 39–47 (2009)
4. Klamka, J.: Stochastic controllability of systems with variable delay in control. Bulletin of the Polish Academy of Sciences. Technical Sciences 56(3), 279–284 (2008)
5. Klamka, J.: Constrained controllability of semilinear systems with delayed controls. Bulletin of the Polish Academy of Sciences. Technical Sciences 56(4), 333–337 (2008)
6. Klamka, J.: Stochastic controllability and minimum energy control of systems with multiple delays in control. Applied Mathematics and Computation 206(2), 704–715 (2008)
7. Klamka, J.: Constrained controllability of semilinear systems with delays. Nonlinear Dynamics 56(1-2), 169–177 (2009)
8. Klamka, J.: Stochastic controllability of systems with multiple delays in control. International Journal of Applied Mathematics and Computer Science 19(1), 39–47 (2009)

Chapter 2
Comparison of Rules Synthesis Methods Accuracy in the System of Type 1 Diabetes Prediction

Rafal Deja

Abstract. While creating the decision support system we encounter the classification accuracy problem. In the paper author compares the accuracy of two rules synthesis algorithms based on the rough set theory. This comparison is based on the medical support system that goal is to predict the illness among the children with genetic susceptibility to DMT1. The system can help to recommend including a person to pre-diabetes therapy.

2.1 Introduction

It is well known that type 1 diabetes is a disease with the genetic background. Many authors [3] have shown that the risk of developing the disease is much higher among the first degree relatives than in the population. At the same time the above-quoted results mean that not all children with genetic susceptibility to DMT1 will ever develop the disease. Epidemiological data provide evidence to link some environmental triggers with DMT1 (e.g. viruses). In this study the genes constituting the HLA complex has been considered. It is believed [3] the genes from HLA complex become the most important genetic factor of susceptibility to DMT1. Additionally the genetic predisposition of Th1 and Th2 lymphocytes cytokine's production can introduce the onset of the illness, thus these genetic results has been consider either.

We developed the information system that allows the classification of children with genetic susceptibility to DMT1 to those with higher and lower risk of falling ill [4]. The children from the first category can be treated with pre-diabetes therapy by doctors. The aim of the study is, while creating the decision support system, to measure and possibly improve the classification performance.

Rafal Deja
Academy of Business in Dabrowa Gornicza, Department of Computer Science
ul. Cieplaka 1c, 41-300 Dabrowa Gornicza, Poland
e-mail: rdeja@wsb.edu.pl
http://www.wsb.edu.pl

A. Kapczynski et al. (Eds.): Internet - Technical Develop. & Appli. 2, AISC 118, pp. 13–24.
springerlink.com

The data collected in the research like other medical data are complex, sometimes uncertain and incomplete. Furthermore the relations among the attributes are hard to be described. Thus we decided to use algorithms and methods based on the rough set theory. In many papers [8,13,14,15,16] the suitability of these methods in medical analysis has been proven.

2.2 Data Description

In the study we collected the data of

1. 44 children with diabetes mellitus type 1
2. 44 healthy siblings of the ill children
3. 36 persons without any symptoms of illness and genetic predisposition in medical history (healthy control group).

The attributes considered in the study are presented in Table 2.1.

Table 2.1 The attributes used in the study

Attribute (A)	Card. of value set of A	Medical meaning
DRB1	24	Allele of class II subregion of DRB of the HLA complex on the first chromosome
DRB2	24	Allele of class II subregion of DRB of the HLA complex on the second chromosome
DQB1	12	Allele of class II subregion DQB of the HLA complex on the first chromosome
DQB2	13	Allele of class II subregion DQB of the HLA complex on the second chromosome
TNF	2	TNF-alpha gene -308 A/G polymorphism on the first chromosome
TNF2	2	TNF-alpha gene -308 A/G polymorphism on the second chromosome
IL10	4	Interleukin-10 gene polymorphism on the first chromosome
IL102	3	Interleukin-10 gene polymorphism on the second chromosome
IL6	2	Interleukin-6 gene polymorphism on the first chromosome
IL62	2	Interleukin-6 gene polymorphism on the second chromosome
IFN	2	Interferon gene polymorphism on the first chromosome
IFN2	2	Interferon gene polymorphism on the second chromosome

Table 2.2 Decision table DMT1 children vs. healthy sibling (fragment)

DRB1	DQB1	TNF	IL10	IL6	IFN	DRB2	DQB2	INF	IL102	IL62	IFN2	D
301	201	A	ATA	G	T	1601	501	A	ATA	G	T	1
701	201	G	GCC	G	T	301	201	G	ACC	C	A	1
1601	502	G	ACC	G	T	401	302	A	ACC	C	T	1
401	302	G	GCC	G	T	301	201	A	ACC	C	A	1
701	202	G	GCC	C	T	404	302	G	ATA	C	T	1
1502	602	G	GCC	C	A	1601	501	G	GCC	C	A	0
701	201	G	GCC	G	T	701	201	G	GCC	C	A	0

Based on the data three decision tables can be created. The first one consists of the sick children and their healthy siblings' results of medical examinations. The second one consists of the data of sick children and the control group and the third table gathers the healthy siblings and the control group data. The exemplar decision table of children with diabetes and their healthy siblings is presented in Table 2.2. The decision D=1 denotes ill children and D=0 denotes healthy sibling.

2.3 Problem Description

The classification problem consists on dividing objects into disjoint classes, which form the partition of the universe. Let assume the existence of the unique attribute D, which denotes the class and is taking the class number as a value. Having the information table $I=(A, V_a)$ the goal is to find classification rules of the form, $r=>D=d_i$, where r is the formula over the A and $d_i$is the class number.

In typical scenario of solving the classification problem [1] the input data set is divided into two separated parts. Using one of the parts, called the training set, the classifier is taught. The classifier is applied and verified using the remaining part, called the test set. The result of the process is the classification of the test set and the structure of the classifier that can be used as a decision system. This method is often called *train-and-test*.

In the paper the main problem is to classify the children with susceptibility to DMT1 in to those with higher and lower risk of getting ill. The starting point is the classifier of being sick. We obtain this classifier by inferring rules from the decision table children with diabetes vs. healthy control group. The classifier of being sick is going to be applied into the table of healthy sibling group to classify them into those with lower and higher risk (diabetes prediction classifier). The process is illustrated with Figure 2.1. The classification process yields in generating the decision support system. The crucial question arises about the quality of the system. In the paper the quality of the classifiers obtained with different methods is compared and discussed.

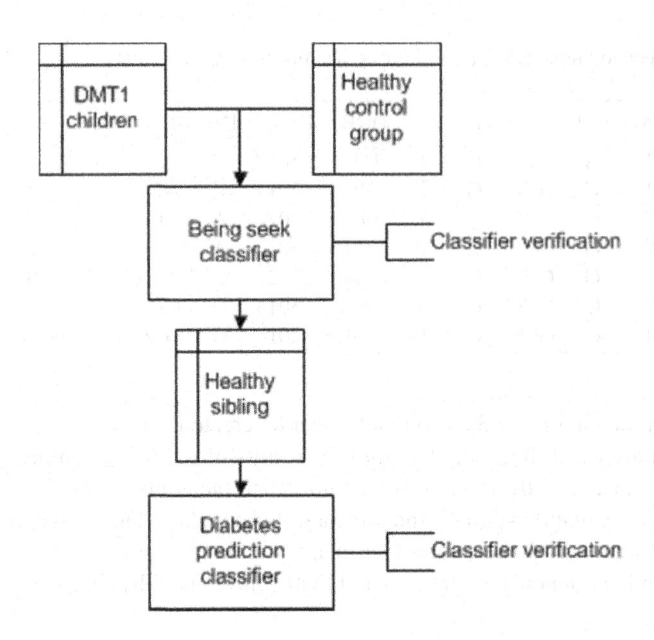

Fig. 2.1 The classification process

2.4 Methods

2.4.1 Rough Set Preliminaries

Rough set theory was developed by Zdzislaw Pawlak [11] in the early 1980's. It is an important tool to deal with uncertain or vague knowledge and allows clear classificatory analysis of data tables.

An information system is a pair $S=(U, A)$, where U - is nonempty, finite set called the universe; elements of U are called objects, A - is nonempty, finite set of attributes. Every attribute $a \in A$ is a map, $a: U \to V_a$, where the set V_a is the value set of a. A decision system is any information system of the form $I=(U, A \cup \{d\})$, where $d \notin A$ is called the decision attribute. The elements of A are called conditional attributes. With any not empty set $B \subseteq A$, we define the equivalence relation called B-indiscernibility: $IND(B) = \{(x,y) \in U \times U : \forall a \in B\,(a(x) = a(y))\}$. Objects x, y that belong to the relation $IND(B)$ are indiscernible from each other by attributes from B. The minimal subset $B \subseteq A$ such that $IND(A)=IND(B)$ is called the reduct of set A. The set of all reducts from A is denoted by $RED(A)$. The relation $IND(B)$ introduces the division of all objects from U into equivalence classes $[x]_B = \{y : xIND(B)y\}$ for each $x \in U$. The sets $\underline{B}X = \{x \in U : [x]_B \subseteq X\}$, $\overline{B}X = \{x \in U : [x]_B \bigcap X \neq \emptyset\}$ are called the B–lower and B–upper approximations of X, respectively. The set $\underline{B}X$ is also called positive region of X. The accuracy of approximation can be measured with coefficient $\propto_B(X) = \frac{|\underline{B}X|}{|\overline{B}X|}$, where $|X|$denotes the cardinality of X. Obviously 0

$<= \propto_B (X) <= 1$. If $\propto_B (X)=1$, X is crisp with respect to B, and otherwise, if $\propto_B (X) < 1$, X is rough with respect to B.

The subsystem of the information system I is an information system I' with the same attributes and reduce set of objects i.e. if $I=(U, A)$ then $I'=(U', A')$, where $x' \subseteq U$, $A'=\{a{:}a \in A\}$. In the set of all reducts RED_I for a given information system the subset of attributes can be distinguished that is not only the reduct of I but also the reduct of family of subsystems of I. A dynamic reduct of I is a reduct of I that is also a reduct of some (as much as required) subsystems of I. Formally the quality of the dynamic reduct B is linked with the number of subsystems B is reduct for, in the following way: $Stab(B,F) = \frac{|I' \in F: B \in RED_{I}'|}{|F|}$, where $Stab(B, F)$ is called stability coefficient of dynamic reduct B.

2.4.2 Rules Synthesis

2.4.2.1 Rules Synthesis Using Reducts

The straight forward rules generation [9], that yields the minimal decision rules with respect to the number of attributes used at the left side, consists of the following steps:

1. Building the discernibility matrix (and discernibility function)
2. Calculating the reducts
3. Constructing rules by overlaying the reducts to the originating decision table.

In the paper for rules generation the dynamic reducts concept will be applied. The dynamic reducts classify the new cases better than the entire set of reducts while finding the shortest reduct with respect to the number of attributes is the NP-hard problem [9]. It is also stated [1] that dynamic reducts reflects some global properties, independent of the particular (randomly chosen) training set. For rules generation the dynamic reducts with stability coefficient greater than 0.5 are considered.

2.4.2.2 LEM2 Algorithm

The other algorithm of rules synthesis considered in the paper is optimized in producing the minimal set of rules. The LEM2 (Learning from Examples, Module 2) algorithm has originally been proposed by Grzymala-Busse [5, 6] and has been successfully applied in many medical problems [8, 13, 16]. The main ideas of the algorithm are as follows [7]. A block of an attribute-value pair t =(a, v), denoted [t], is the set of all examples that for attribute a have value v. A concept, described by the value w of decision d, is denoted [(d, w)], and it is the set of all examples that have value w for decision d. Let C be a concept and let T be a set of attribute-value pairs. Concept C depends on a set T if and only if $\emptyset \neq [T] = \bigcap_{t \in T} [t] \subseteq C$ Set T is a minimal complex of concept C if and only if C depends on T and T is minimal. Let **T** be a nonempty collection of nonempty sets of attribute-value pairs. Set **T** is a local covering of C if and only if the following three conditions are satisfied:

1. each member of **T** is a minimal complex of C,
2. $\bigcup_{T \in T} [T] = B$, and
3. **T** is minimal, i.e., **T** has the smallest possible number of members.

For each concept C, the LEM2 algorithm induces rules by computing a local covering **T**. Any set T, a minimal complex which is a member of **T**, is computed from attribute-value pairs selected from the set T(G) of attribute-value pairs relevant with a current goal G, i.e., pairs whose blocks have nonempty intersection with G. The initial goal G is equal to the concept and then it is iteratively updated by subtracting from G the set of examples described by the set of minimal complexes computed so far. Attribute-value pairs from T which are selected as the most relevant, i.e., on the basis of maximum of the cardinality of $[t] \cup G$, if a tie occurs, on the basis of the smallest cardinality of [t]. The last condition is equivalent to the maximal conditional probability of goal G given attribute-value pair t. The coverage parameter for LEM2 algorithm has been set to 0.9 in all experiments.

2.4.3 Classification Accuracy

The standard way to verify the quality of the classifier is to apply it to the test set (the classification of which is known) and compare the original classification with the new one. There still remains the problem of dividing the data set into the training part and the test set. The most common way is to randomly divide the data set into several parts. Then during evaluation, all but one part are used to generate the model and the remaining one to evaluate the model. This process is repeated for every part of the original set. Usually (because the quality of the model improves with the size of the training set) the number of parts is big and the quantity of cases in each part is small. That approach is called N-folded cross validation where one Nth of the table is left out of the learning process and later used for testing.

We divided the decision table children with diabetes vs. healthy control group into 10 training and test sets by randomly choosing about 10% of healthy and about 10% of sick children. The computer program for experiments with classifiers validation has been prepared by the author.

The accuracy coefficient of the pair training set/test set t is calculated in the following way:

$$C_t = \frac{\sum_{i=1..n} \gamma_i}{|X_t|} \cdot 100\%$$

and γ_i is equal: 1 – for correct classification of object i from test set t, 0 – wrong classification, 0.5 – decision is missing – the ith object cannot be surely classified. The $|X_t|$ denotes the cardinality of the test set corresponding to the training set t.

We are going to apply the algorithm described above to the system of Type 1 diabetes prediction to verify the quality of the being sick classifier. The next step is to infer the classifier of being more risky to fall ill while the children with susceptibility to DMT1 are considered (diabetes prediction classifier). The cross validation approach to verify the diabetes prediction classifier cannot be applied because the classification of the healthy sibling set is not given (new cases). The proposal of the solution is to verify the quality of the classification by asking the expert. We can't ask the expert whether the classification is good (the new cases are also new for the expert) thus the idea is to infer the classifier from the new cases and ask the expert about the classifier (there are other researches/believes that can be confronted with the results of this work).

2.5 Results

2.5.1 RSES System

Most of the analysis and experiments were conducted using RSES (Rough Set Exploration System) system [2]. The system (current version is 2.2) has been created by so called Group of Logic at the Warsaw University http://logic.mimuw.edu.pl/ under the supervision of professor Andrzej Skowron.

2.5.2 Being Sick Classifier

For each training set the algorithm of rules synthesis by using dynamic reducts and LEM2 is applied and verified against the corresponding test set. The results of classification accuracy have been presented in Table 2.3.

Table 2.3 The accuracy coefficient of being sick classifier using reducts

Training set	$\mathfrak{C}^{RED}$	$\mathfrak{C}^{LEM2}$
T_1	87.5	93.75
T_2	93.75	81.25
T_3	68.75	75
T_4	93.75	93.75
T_5	87.5	68.75
T_6	81.25	75
T_7	100	93.75
T_8	87.5	93.75
T_9	87.5	93.75
T_{10}	93.75	100
Average	88.13	86.88

The average classification accuracy using both algorithms is nearly the same, with a slight advantage for dynamic reducts.

2.5.2.1 Results Using Reducts

The example results for Training set 7 are as follows: there are 19 reducts, and 874 rules inferred. The examples of reducts and rules are presented below. The decision value D=0 means healthy and D=1 means sick.

RED(A) = { { DR, TNF, IL10, DR2, IL102 }, { DR, DQ, TNF, DR2, IL102 }, { DR, DQ, TNF, IL6, TNF2, IL102, IL62, IFN2 }, { DR, DQ, TNF, IL10, TNF2, IL102, IL62, IFN2 }, { DR, DQ, TNF, DR2, IL62 }... }

The examples of rules generated from reducts:

(DR=301)&(TNF=A)&(IL10=ATA)&(DR2=1601)&(IL102=ATA)=>(D=1)
(DR=701)&(TNF=G)&(IL10=GCC)&(DR2=301)&(IL102=ACC)=>(D=1)
(DR=1601)&(TNF=G)&(IL10=ACC)&(DR2=401)&(IL102=ACC)=>(D=1)
(DR=301)&(TNF=G)&(IL10=GCC)&(DR2=701)&(IL102=ATA)=>(D=0)
(DR=401)&(TNF=G)&(IL10=GCC)&(DR2=1103)&(IL102=ATA)=>(D=0)
(DR=1301)&(TNF=G)&(IL10=GCC)&(DR2=1301)&(IL102=ACC)=>(D=0)
(DR=1301)&(TNF=G)&(IL10=GCC)&(DR2=102)&(IL102=GCC)=>(D=0)
....

2.5.2.2 Results Using LEM2 Algorithm

The example results for training set 10 of rules synthesis using LEM 2 algorithm are partially presented below.

(TNF=G)&(IFN=T)&(IL6=G)&(IL62=G)&(IFN2=A)=>(D=0[11])
(TNF=G)&(IL62=C)&(IFN2=A)&(TNF2=A)&(IFN=T)&(IL10=GCC)
&(DQ2=201)=>(D=1[7])
(TNF=G)&(IL62=C)&(IFN2=A)&(IL10=ACC)&(IL6=C)=>(D=1[6])
(TNF=G)&(IFN=T)&(TNF2=G)&(IFN2=T)&(DR=1501)=>(D=0[6])
(TNF=G)&(IFN=T)&(IL6=G)&(IFN2=A)&(IL10=GCC)&(TNF2=G)
&(DQ=301)=>(D=0[5])
(IL6=G)&(IL10=GCC)&(DQ=201)&(TNF2=A)&(DR=301)&(IFN2=A)
&(TNF=A)=>(D=1[4])
(IFN=T)&(IL10=GCC)&(IL62=C)&(IL102=ATA)&(IL6=C)=>(D=1[4])
(IL6=G)&(IFN=T)&(IFN2=T)&(DQ=201)=>(D=1[4])
(IL6=G)&(TNF=G)&(IFN=T)&(IL102=ATA)&(IL10=ACC)
&(DR=301)=>(D=1[3])
(TNF=G)&(IL6=G)&(TNF2=G)&(IL102=ATA)&(IL62=G)&(IL10=ACC)
&(DQ2=201)=>(D=1[3])
...

There are 23 rules inferred. The decision value D=0 means healthy and D=1 means sick. In the square brackets the power of each rule is provided.

2.5.3 Diabetes Prediction Classifier

For both algorithms the whole decision table TDM1 children vs. control group has been taken again and the being seek classifier has been inferred. Applying obtained rules to the healthy sibling table we can classify the children from that table to those potentially sick and potentially healthy. The partial results using LEM2 algorithm is presented in Table 2.4.

Table 2.4 Decision table healthy sibling with classification (fragment)

DRB1	DQB1	TNF	IL10	IL6	IFN	DRB2	DQB2	INF2	IL102	IL62	IFN2	D
1502	602	G	GCC	C	A	1601	501	G	GCC	C	A	MISSING
701	201	G	GCC	G	T	701	201	G	GCC	C	A	1
301	201	G	ACC	G	A	402	302	A	ACC	C	A	MISSING
1601	502	G	ACC	G	T	401	302	A	ACC	C	T	MISSING
401	302	G	GCC	G	T	301	201	A	ACC	C	A	1
301	201	G	GCC	G	T	404	302	A	ACC	C	T	1
101	501	G	GCC	C	T	419	304	G	ATA	C	A	1
404	302	G	GCC	G	A	1103	301	G	ATA	C	A	1
401	302	G	GCC	G	T	1601	501	G	ACC	C	T	MISSING
701	202	G	GCC	G	T	402	302	G	ACC	C	T	1
101	501	G	ACC	G	T	401	302	G	ATA	G	A	0
1306	602	A	GCC	C	T	401	302	A	ATA	C	A	1
1601	502	G	GCC	G	A	101	501	A	ACC	G	A	MISSING

The decision table above (2.4 - classified healthy siblings) is the basis of the Type 1 diabetes prediction system. The rules generated from the table above can be treated as a decision support system.

2.5.3.1 Results Using Dynamic Reducts

Using dynamic reducts we obtain 2 reducts and 81 rules (included rules indicated unclassified objects). Some of the rules are as follows (decision D=1 means the prediction of being sick and D=0 of being healthy, the number in square brackets is the power of the rule):

(DR=701)&(DR2=701)&(IFN2=A)=>(D=0[1])
(DR=1501)&(DR2=301)&(IFN2=A)=>(D=0[2])
(DR=1601)&(DR2=401)&(IFN2=T)=>(D=1[1])
(DR=401)&(DR2=301)&(IFN2=A)=>(D=1[2])
(DR=101)&(DR2=419)&(IFN2=A)=>(D=1[1])
(DR=301)&(DR2=401)&(IFN2=A)=>(D=1[2])
(IL10=GCC)&(IL6=G)&(DR2=701)&(IL102=GCC)&(IFN2=A)=>(D=0[2])
(IL10=GCC)&(IL6=G)&(DR2=701)&(IL102=ATA)&(IFN2=A)=>(D=1[2])
...

2.5.3.2 Results Using LEM2 Algorithm

When applying LEM2 algorithm there are 13 rules: 3 of them classify children to potentially healthy, 4 to potentially sick, remaining rules indicates the gens arrangement that cannot be surely classified. The rules are presented below:

Potentially healthy children:

(TNF=G)&(IL10=GCC)&(IFN=A)&(IL62=C)&(IFN2=A)&(IL6=C)
&(DR=1502)=>(D=0[1])

(TNF=G)&(IL10=GCC)&(IFN=A)&(TNF2=A)&(IL102=ATA)=>(D=0[2])
(TNF=G)&(IFN2=A)&(IL102=ATA)&(IL6=G)&(IL62=G)=>(D=0[4])

Potentially sick children:

(IL62=C)&(IFN=T)&(TNF2=A)&(IFN2=A)&(TNF=G)
&(IL10=ACC)=>(D=1[4])
(IL62=C)&(TNF=G)&(IL6=G)&(IL10=GCC)&(TNF2=G)=>(D=1[9])
(IL62=C)&(IFN=T)&(DR2=401)&(DQ2=302)&(TNF2=A)&(TNF=A)
&(IL10=GCC)=>(D=1[2])
(TNF=G)&(IL62=C)&(IFN=T)&(DR=1601)=>(D=1[2])

2.5.4 Diabetes Prediction Classification Accuracy

The classification accuracy of being more risky to fall ill cannot be verified using the N-folded cross validation - the classification of the healthy sibling set is not given (new cases). The idea was to verify the quality of the classification by asking the expert.

2.5.4.1 Expert Evaluation

In the medical papers (researches) the statistical methods are usually used for measuring the influence of selected factors to susceptibility to DMT1. Either the attributes are compared one by one looking for the statistical significance or the multiple logistic regressions is used to build the model using some attributes looking for those with high odds ratio. Unfortunately there are no researches which results can be directly compared with the paper.

According to expert knowledge the complex DR=301/DQ=302 indicates the high risk of falling ill. The other predictors of diabetes are alleles DR=40x, DQ=201 and TNF=A. The polymorphism of cytokines genes is not statistically relevant to develop the disease according to expert knowledge. When analyzing the rules obtained one can find the following conclusions. There exists one rule among those obtained using LEM2 algorithm, and 6 when using dynamic reducts with conditions corresponding to the expert knowledge, however the number of LEM2 reducts is significantly smaller (4 against 38). There also must be remembered that the rule (from the prediction system above) is fired only when all conditions are satisfied. Therefore one can find some rules obtained using dynamic reducts more relevant than using LEM2 algorithm.

2.5.4.2 Cross Validation

There are several theories while the siblings of ill children are still healthy. One of the ideas suggests that some of the healthy children didn't encounter the environment trigger yet, but their genes arrangement is adequate to ill children. One can imagine that the diabetes prediction classifier should classify the being seek concept well. Thus we can perform cross validation of our prediction classifier using sick children and the control group information table. Using the same accuracy coefficient as defined in section 2.3 we obtained the following results: $\complement^{RED}$=63.75, $\complement^{LEM2}$=53.125. The results are not satisfactory - the diabetes prediction classifier is quite different than being sick classifier. One can interpret the results with the conclusions that there exists the gens arrangement that distinguishes the potentially healthy and potentially sick children among the children with genetic susceptibility to DMT1.

2.6 Conclusions

In the paper author has presented two algorithms for rules synthesis based on the rough set theory; direct rules generation using dynamic reducts and the LEM2 approach. Those algorithm has been chosen as those discussed in many papers as of the first choice in machine learning.

The main goal of the experiments was to develop the expert system that allows the classification of children with genetic susceptibility to DMT1 to those with higher and lower risk of falling ill and verify the rules quality. The goal has been reached with two steps: developing the classifier of being sick and based of it developing the diabetes prediction classifier in the second step. The classifier from the first step has been verified using the N-folded cross validation approach and the second one by asking the expert. The average accuracy of classifying new cases using the direct rules generation (88.13%) was slightly better than the result using LEM2 algorithm (86.875%). However, using the LEM2 algorithm the number of rules inferred is much less than using straight forward algorithm. We could have expected that the LEM2 algorithm will classify new cases better. The power of the rules inferred with LEM2 algorithm is bigger. In evaluating the classification algorithms one can experiment with many parameters, which here were taken arbitrary based on the author knowledge.

The results of LEM2 algorithm are more suitable for understanding, use and verification by physicians. According to expert knowledge the diabetes prediction classifier obtained both by applying LEM2 algorithm and using dynamic reducts are useful in medical practice however no relevant researches have been conducted, that can confirm the results.

Acknowledgements. Data has been obtained from the Pediatric, Endocrinology and Diabetology Department of Silesian Medical University thanks to courtesy of Grazyna Deja, MD, PhD.

References

1. Bazan, J., Nguyen, H.S., Nguyen, S.H., Synak, P., Wróblewski, J.: Rough set algorithms in classification problems. In: Polkowski, L., Lin, T.Y., Tsumoto, S. (eds.) Rough Set Methods and Applications: New Developments in Knowledge Discovery in Information Systems. Studies in Fuzziness and Soft Computing, vol. 56, pp. 49–88. Physica-Verlag, Heidelberg (2000)

2. Bazan, J.G., Szczuka, M.S., Wróblewski, J.: A New Version of Rough Set Exploration System. In: Alpigini, J.J., Peters, J.F., Skowron, A., Zhong, N. (eds.) RSCTC 2002. LNCS (LNAI), vol. 2475, pp. 397–404. Springer, Heidelberg (2002)
3. Deja, G., Jarosz-Chobot, P., Polañska, J., Siekiera, U., Maşecka-Tendera, E.: Is the association between TNF-alpha-308 A allele and DMT1 independent of HLA-DRB1, DQB1 alleles? Mediators Inflamm. 2006, 19724 (2006)
4. Deja, R.: Applying rough set theory to the system of type 1 diabetes prediction. In: Tkacz, E., Kapczynski, A. (eds.) Internet – Technical Development and Applications. Advances in Intelligent and Soft Computing, vol. 64, pp. 119–127. Springer, Heidelberg (2009)
5. Grzymala-Busse, J., Wang, A.: Modified algorithms lem1 and lem2 for rule induction from data with missing attribute values. In: Proc. of 5th Int. Workshop on Rough Sets and Soft Computing, pp. 69–72 (1997)
6. Grzymala-Busse, J.W.: Mlem2-discretization during rule induction. In: Proceedings of the International IIS, pp. 499–508 (2003)
7. Grzymala-Busse, J.W.: Selected Algorithms of Machine Learning from Examples. Fundamenta Informaticae 18, 193–207 (1993)
8. Ilczuk, G., Wakulicz-Deja, A.: Rough sets approach to medical diagnosis system. In: Szczepaniak, P.S., Kacprzyk, J., Niewiadomski, A. (eds.) AWIC 2005. LNCS (LNAI), vol. 3528, pp. 204–210. Springer, Heidelberg (2005)
9. Komorowski, H.J., Pawlak, Z., Polkowski, L.T., Skowron, A.: Rough Sets: A Tutorial, pp. 3–98. Springer, Singapore (1999)
10. Midelfart, H., Komorowski, H.J., Norsett, K.G., Yadetie, F., Sandvik, A.K., Laegreid, A.: Learning rough set classifiers from gene expressions and clinical data. Fundamenta Informaticae 53, 155–183 (2002)
11. Pawlak, Z.: Rough Sets: Theoretical aspects of reasoning about data. Kluwer Academic Publishers, Boston (1991)
12. Skowron, A., Rauszer, G.: The discernibility matrices and functions in information systems. In: Sşowinski, R. (ed.) Intelligent Decision Support. Handbook of Applications and Advances of the Rough Sets Theory, pp. 331–336. Kluwer Academic Publishers (1992)
13. Slowinski, K., Stefanowsk, J., Siwinski, R.: Application of rule induction and rough sets to verification of magnetic resonance diagnosis. Fundam. Inform. 53, 345–363 (2002)
14. Tsumoto, S.: Extracting structure of medical diagnosis: Rough set approach. In Wang, G., Liu, Q., Yao, Y., Skowron, A., eds.: Rough Sets, Fuzzy Sets, Data Mining, and Granular Computing. In: Wang, G., Liu, Q., Yao, Y., Skowron, A. (eds.) RSFDGrC 2003. LNCS (LNAI), vol. 2639, pp. 78–88. Springer, Heidelberg (2003)
15. Tsumoto, S.: Mining diagnostic rules from clinical databases using rough sets and medical diagnostic model. Information Sciences: An International Journal 162, 65–80 (2004)
16. Wakulicz-Deja, A., Paszek, P.: Applying rough set theory to multi stage medical diagnosing. Fundamenta Informaticae 54, 387–408 (2003)

Chapter 3
Progressive 3D Mesh Transmission in IP Networks: Implementation and Traffic Measurements

Slawomir Nowak and Damian Sobczak

Abstract. This work concerns the problem of preparation of the computer system to represent the 3D scene in a dynamic network environment. The article presents preliminary results of network measurement of 3D mesh progressive transmission using TCP protocol. The objective is to evaluate the possible impact of transmission dynamics on the amount of data available in VR browsing client, which relates to browsing Quality of Experience. The article compares results obtained in a real network to the simulation studies, conducted previously. In four sections, the article presents a brief introduction and exploration model, simulation results, implementation of progression and simple network protocol and finally conclusions and future works.

3.1 Introduction

Virtual Reality (VR) is a term that applies to techniques that use computers to build, present, and allow for user interaction with simulated environment. Furthermore, VR covers remote communication which provide virtual presence of observers in a shared environment. It is therefore a natural medium for entertainment, education, data visualization, telemedicine, telepresence, telexistence etc.

Slawomir Nowak
Academy of Business in Dabrowa Gornicza, Department of Computer Science
ul. Cieplaka 1c, 41-300 Dabrowa Gornicza, Poland
e-mail: emanuel@iitis.pl
http://www.wsb.edu.pl

Damian Sobczak
Institute of Theoretical and Applied Informatics
Polish Academy of Sciences
Baltycka 5, 44–100 Gliwice, Poland
e-mail: sobczak.d@gmail.com

A. Kapczynski et al. (Eds.): Internet - Technical Develop. & Appli. 2, AISC 118, pp. 25–37.
springerlink.com

The popularity of VR is growing steadily over the past years with improvement of hardware capabilities allowing for better Quality of (user) Experience (QoE), understood as a subjective measure of a customer's experiences with a service.

The realization of VR service usually requires, beyond traditional multimedia data types, 3D object description, commonly being mesh of vertices [1]. This data type can occupy a substantial portion of the whole VR data.

3.1.1 3D Objects' Transmission

If the data needed to create and maintain a 3D scene are stored locally, the workstation requires appropriate hardware and software resources to represent a 3D scene. In this case, the network interactions, if any, are limited to exchange of information on the mutual position of objects able to interact. When 3D objects must be transmitted from one or more remote network location the problem of providing the required level of QoE is becoming more complex [2].

From network perspective, the process of interaction in VR is quite different than video/audio transmission. While both share the necessity of adapting to network conditions (i.e. by using scalable streams [1] and rate distortion optimization), the VR interactions are dynamic by nature, as it is the user who "decides" what part of the content is to be send over the network. This makes tasks such as traffic prediction, congestion control and buffering difficult.

The problem for the interactive network transmission of 3D objects occurs in many applications. As an example, a remote data repository (such as medical or archeological sites) of very detailed meshes, available as needed, at different levels of detail. Another example is the use of a "virtual museum" where many objects are stored on multiple servers creating space for exploration.

A separate field of applications are applications for so-called *thin clients* (PDAs, smart phones, mobile phones) with very limited local hardware resources. Their functionality definition is based directly on the use of network resources. In case the data must be sent using the limited network resources (specific of public Internet) traffic management for the most efficient use of resources to ensure a satisfactory level of QoE becomes very complex. Even if the network allows to specify the level of quality of service (QoS), a reference to the resulting QoE is also not simple and requires separate research.

In the simplest case (which concerns the implementation and the research presented in this article) we are dealing with communication in the model one client - one server (and one observer). The solution is reduced to progressive transfer of data to the client in the least possible degree of involvement of network resources while ensuring a satisfactory level of QoE observer. One possibility for achieving this goal is the selection of levels of progression depending on the distance of objects from the observer. The problem becomes much more complex in the model in which there is a number of clients (many observers), and many servers. The decision, which parts of the progressive representation of objects should be sent to customers, while maintaining a relatively high level of QoE is not trivial. To solve this problem several additional solution can be considered, e.g. the prediction of observer movement, the scheduling of objects depending on observers positions including the line of sight of the observer etc. They will be the subject of further research.

The next section presents the simulation studies on traffic-related transmissions of 3D progressive meshes and work efforts to obtain a generator for this type of traffic. Then the implementation of Hoppe's progression and simple network protocol is presented. Next the measurements in the TCP/IP network are presented and finally conclusions and future works.

Out studies used TCP to communicate between processes. This approach may cause undesired delay in model rendering (overhead of the connection-like protocol). The potential solution is to use UDP protocol, but the progression in case of some subsequent data are lost the process of refinement must be suspended until the retransmission. However the problem of data loss has been omitted at that stage. The example solutions for that problem is presented e.g. in [5],[6], using hybrid TCP and UDP connection, and appropriate assigning data to TCP or UDP. Another solution (defined as *model-based*) is to prepare error resilient packetization scheme, based on the modified progressive data structure [7].

3.2 Simulation Evaluation

As the quantity of 3D data is expected to grow, it is of importance to analyze the properties of its transmission. The key element for this is an accurate model of 3D scene exploration and related network traffic. Unfortunately, due to many different used technologies and applications, formulation of such model is difficult. To the best of authors' knowledge, there is no universally accepted model or protocol of 3D data transmission yet.

The following subsection describes more detail only the simplest case of exploration, "one object, one observer, one server", as directly related to the scope of implementation, presented in the next chapter.

3.2.1 Simulation Model and Simulation Results

As a result of previous work the simulation model of VR browsing using TCP/IP network was developed [15]. The model uses a client-server architecture. The role of the server is to store 3D objects (meshes), keep a history of data transmitted to a client, and respond to client's request by sending data updates. The role of the client is to present (render) the 3D data, and send positions changes of the virtual observer to the server. Based on the information about the positions of all observers, the server decides to send updates.

Each 3D mesh has its own location (x,y) in the virtual world and is stored in progressive form. It allows to send an incomplete object (with only a small subset of vertices/triangles) and then update it when needed. The controlling parameter is the vertices/triangles density per display unit, as it directly relates to the perceived quality of object's rendering (an important part of overall QoE). When user's position in the virtual world changes, so does the distance to the object. If the distance decreases, the object will occupy a larger portion of client's rendering window. In that case, to sustain the quality, an update of object data must be send, to keep the displayed density on required level.

For single object the relation between the observer's distance to object (in meters) and size (in bytes) is expressed by (3.1):

$$size = \begin{cases} 0,\ (\text{distance} > \text{max distance}) \\ \text{max size},\ (\text{distance} <= \text{min distance}) \\ \text{min size} + \frac{\text{object size}}{\text{distance}^2},\ (\text{mindistance} < \text{distance} <= \text{max distance}) \end{cases} \quad (3.1)$$

The size of subsequent requests for the server of all objects, according to observer's position, is obtained using a dedicated tool, 3DtrafficGenerator [11]. By randomly placing or editing 3D objects and observer's trajectory the configuration file is generated. That file was used as an input for the simulation evaluation.

A simplest case is where there is only one 3D object in the VR space, and the observer is moving closer to that object and no concurrent transmissions in the network. This quantity we denote by hunger_factor (3.2), introduced in [15].

$$hf = \sum_{i=1}^{n} requested\ size(object\ i) - buffered\ data\ size(object\ i) \quad (3.2)$$

It relates to quality loss from the baseline at a given time. Loss of quality takes place after the next movement of the observer in the VR space. It involves the transmission of observers position information to the server. The server sends the update, and loss of quality (measured by hunger_factor) is reduced. The resulting graph is presented on Fig. 3.1.

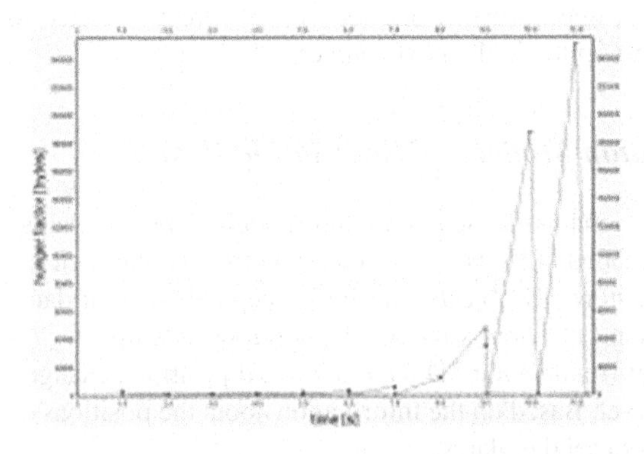

Fig. 3.1 "one object, one observer, one server" case, **hunger_factor** as a function of time.

3.2.2 *Extension of the Simulation Model*

Several different cases of VR exploration were already evaluated. A variety of observers motions have been considered (eg. random Brownian walk and directed random walk) [15]. The simulation model were upgraded with taking into account the viewing angle (sight of view) of the observation. The angle of observation reduces the traffic sent only to the objects remaining in sight of the observer.

Based on the simulation results HMM (Hidden Markov Models) model of traffic generated by 3D server was worked out [14]. Analytical traffic generator can be used both in simulation studies as well as in the real networks.

The simulation results indicate that the nature of 3D traffic is complex, and therefore advanced models (complex network topologies, complex traffic patterns and another exploration models) are needed for performance analysis.

3.3 Implementation

In this chapter the implementation and its scope is described:

1. The server application, which includes Hoppe's progressive mesh decomposition method and prepares data for transmission over the network;
2. The client application, which reproduces data in its progressive form;
3. Implementation of network protocol, description of data serialization and selection method of progression thresholds.

The data preparation for progressive transmission (server-side) and transmission protocol are presented in more detail (the TCP stream used to send subsequent mesh updates, preparation of distance thresholds, additional information necessary to send and allow data serialization, etc.).

The separate problem will be progressive representation and transmission of textures, which plays an important role to achieve the final effect on the level of QoE, however it is not the subject of research covered with this article.

3.3.1 *Preparation of Transmission Data*

The server application uses Hoppe's progression method of three-dimensional meshes [9] modified by Stan Melax [12]. The modification applies to algorithm of choosing next edge to collapse. Melax proposed method described by below equation:

$$\text{cost}(u,v) = ||u-v|| \times \max_{f \in Tu}\{\min_{n \in Tuv}\{(1 - f.normal \bullet n.normal) \div 2\}\} \quad (3.3)$$

where Tu is the set of triangles that contains vertex u and Tuv is the set of triangles that contains both u and v. Every 3D mesh consists of two arrays: vertices and indices array. The first one contains vertex coordinate data in 3D space. Indices array contains information needed to join vertices into triangles. In progressive mesh both arrays are sorted by reduction algorithm and are reflecting the order in which vertices must be added to mesh to obtain optimal shape of the object represented by this mesh.

The result of mesh reduction algorithm is a set of vertex split records, called vsplit. A single vsplit object contains information about newly added vertex and changes it made in indices array. While the reconstructed mesh receives next vsplit records it adds one vertex to its vertices array and changes its indices array to reflect adding of one or two new triangles.

```
Struct vsplit {
Struct position {
float x;
float y;
float z;
}
Array<int> indices;
Array<int> changedIndicesAdresses;
int changedIndicesValue;
}
```

The server application prepares vsplit stream data and sends it to connected client. While the client receives next vsplit records it rebuilds original mesh. Because of various length of vsplit records the server needs to save and send additional information. Sequence of vsplit records must also be transformed from high abstraction level, like instance of class, to low level of abstraction, like sequence of bytes. When client application will receive byte stream, it must reconstruct it again to high level abstraction of instance of class. So, with vsplit records data, server needs to send additional information which describes data structure.

This problem is not trivial. There is no ready-made solutions in literature related to progressive transmission of meshes. The easiest way would be based on the serialization methods. First is the serialization of instances of vsplit objects using self-describing XML language, or with using of SOAP protocol. It is simplest way, but it's occupied with a big amount of describing data. The single serialized vsplit object could take about 1.5kB of storage. The second way is to use binary serialization, where data are kept in binary form. In this case amount of storage memory needed to save single vsplit record was 700B. This method allowed to save about 50% of storage place in relation to XML serialization. However using binary serialization, transmission of single mesh composed of 10 000 vertices took about 7MB of data, transmitted via TCP network. To send more detailed mesh, built with 100 000 vertices, amount of data to transmit increase to 70MB. To send several objects in the same time, or when the server will send object to more than one client, amount of transmitted data accordingly continue to increase.

To solve this problem a dedicated *Update Mesh Record* (UMRecord) was proposed (Fig. 3.2 presents scheme of data sequence). The data are kept in a byte array. Each vsplit record is preserved in a separate array:

1. First four bytes contains length of whole data structure. This information is needed by client to divide data stream into data sequences.
2. Next 12 bytes contain X, Y and Z coordinates of vertex being added to resulting mesh. According to Hoppe's progression method, when adding one vertex to a

mesh, one, two or none triangles are added in its neighborhood. It means, that three, six or none indices are added to indices array. Every index in indices array occupies 4 bytes. When single vsplit record is applied to a mesh, 0, 12 or 24 bytes are added to resulting mesh's indices array. So next 4 bytes contain number of indices added to a mesh, and next 0 to 24 bytes contain indices. In some cases, when a vertex is added to a mesh, some indices are changing its value. It reflects changes in mesh structure.

3. Number of indices that has changed value are kept in next 4 bytes of sequence. Addresses of this indices are kept in next few bytes. Amount of bytes needed to store addresses depends on number of changed indices.
4. The last 4 bytes contain value to which those indices are changing.

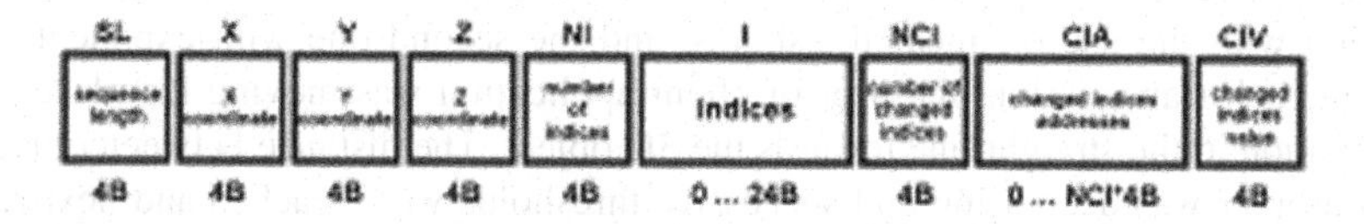

Fig. 3.2 Scheme of UMRecord, used to store single vsplit record.

Minimal amount of data needed to store single vsplit record equals 28 bytes. Maximal amount depends on number of changes in reconstructed mesh'es indices array. This value is increasing with density of original mesh, because more triangles are placed in neighbor of vertices.

Amount of data needed to store single vsplit record oscillates from 28 to 92 bytes. Average volume of data needed to store mesh built with 10 000 vertices is 700kB. It is ten times less than using binary serialization and about twenty times less using XML serialization.

3.3.2 Client-Server Communication Protocol and Thresholds Description

The connection between client and server is set using TCP connection. After simple authorization of client, server waits for requests from client or until client disconnects. If the client sends *update* request, then the server will send some number of UMRecords. The number of UMRecord sent in response depends on *thresholds* calculation. Threshold is a segment of initial distance between an 3D object and observer. For the purposes of experiments the number of thresholds was set arbitrary to 10. Threshold could be set to number within range from one to n, where n is a number of vertices in original mesh (in practical applications the number of thresholds will depend on the properties of the object itself, network performance and QoE requirements).

There are two types of thresholds: *linear* and *exponential*. The amount of vsplit records in each linear thresholds are equal while in exponential type subsequent thresholds contain squarely growing number of vsplits, which reflects the natural process of observation (by reducing the distance, observed surface of objects *grow* in a square).

The research presented in this article concerns two separate transmissions: one with use of linear thresholds and one with use of exponential thresholds.

3.4 3D Traffic Measurement

For measurement of network traffic Wireshark (http://www.wireshark.org) was used. Network traffic measurements was performed during two separate transmissions. Each transmission was sending the same progressive mesh between two computers in simple local area network. The server and the client was connected by single TP-Link WR340G router (100Mb/s interfaces). The first transmission was realized with the use of linear thresholds, and the second one with exponential. During each transmission, the user of client application was moving his place of observation in the straight line towards the 3D object. The distance between object and observer was decreasing and successive thresholds were reached and next request for more progressive data was sent to the server. Server answers by sending an UMRecord, and the client updates the rendered mesh.

Measurement for linear and exponential thresholds and the evaluation of results in terms of QoE are presented and discussed in the next subsection.

3.4.1 Experimental Results

The progressive mesh *sabines.obj*, which has been transmitted, has 53829 vertices and takes 3553356 bytes of memory. This mesh was created by scanning selected items from Museum in Gliwice with 3D scanner [13]. Experiments were performed on other meshes too and the obtained results were similar.

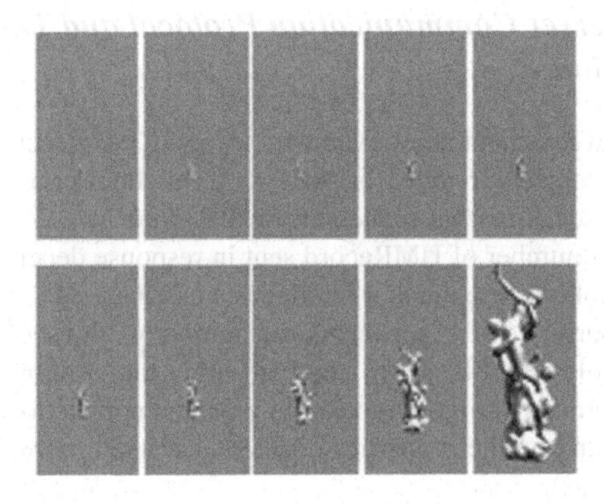

Fig. 3.3 Mesh sabines.obj seen at subsequent linear thresholds.

Fig. 3.4 Visual differences of sabines.obj mesh for subsequent linear thresholds.

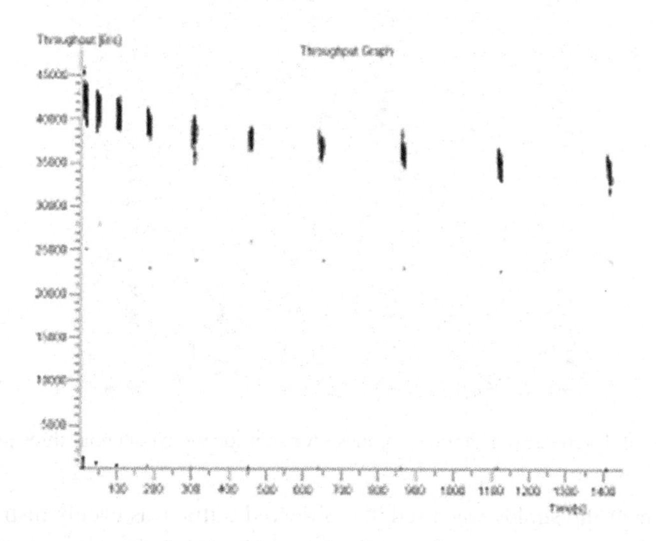

Fig. 3.5 Throughput graph of network traffic observed during transmission of sabines.obj mesh using linear thresholds.

Fig. 3.5 shows throughput (measured on clients-site), which is placed between 32kB/s and 46kB/s. The highest values of throughput are in correlation of time when the subsequent parts of progressive data was sent.

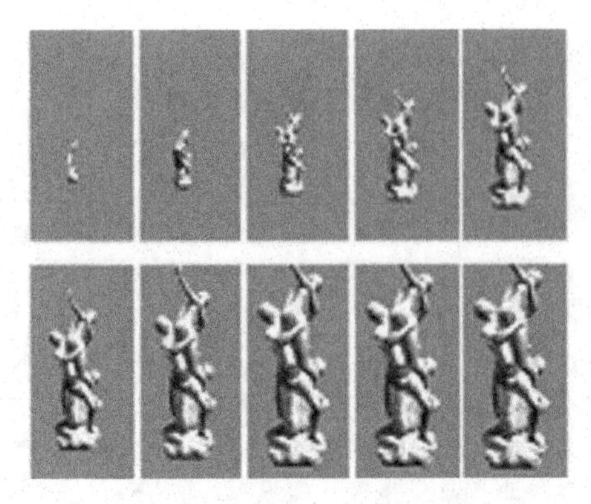

Fig. 3.6 Mesh sabines.obj seen at subsequent exponential thresholds.

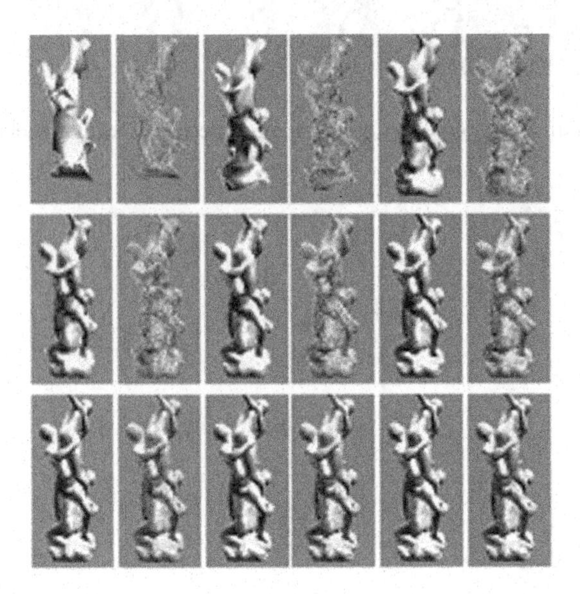

Fig. 3.7 Visual differences of sabines.obj mesh for subsequent exponential thresholds.

When linear thresholds was used, the observed traffic was evenly distributed in time and every stage of transmission sent equal amount of data. It is characteristic that it is hard to see the differences between meshes at high-value thresholds.

In distances far from observer hardly keeps the shape of original mesh (Fig. 3.6) but because of the distance, a substantial inaccuracy of the mesh can remain unnoticed for the observer. Fig. 3.7 shows differences between several levels of details for each threshold seen from constant distance.

Fig. 3.8 shows throughput graph of network traffic, that was observed during transmission of *sabines.obj* mesh. It confirms that for exponential thresholds that are lying far from the observer the network traffic is inconsiderable and sometimes

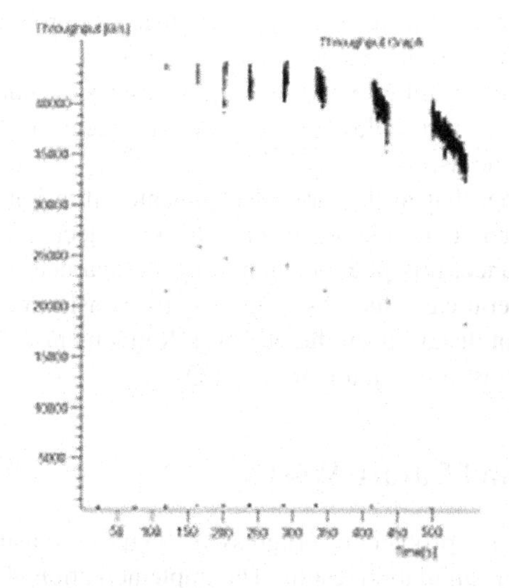

Fig. 3.8 Throughput graph of network traffic observed during transmission of sabines.obj mesh using exponential thresholds.

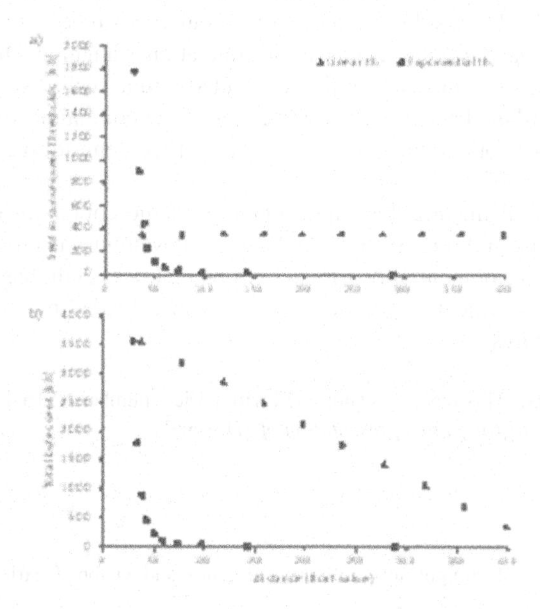

Fig. 3.9 Relation between the 3D object-observer distance and number of bytes sent in subsequent thresholds (a) and total bytes received (b) for linear and exponential thresholds.

even unnoticeable. When the closer to observer distances are reached the network traffic increases. In this case it falls within the range between 32kB/s and 44kB/s.

Fig. 3.9 shows relation between the 3D object-observer distance and number of bytes sent in subsequent thresholds (a) and total bytes received (b) in case of linear and exponential thresholds.

It is worth noticing, that in the case of exponential threshold most of the data is transmitted (data burst) in a short distance to the object. From the perspective of network performance, this phenomenon may be regarded as disadvantageous. However, in the general case, the observation space consists of many 3D objects, most of whom remain distant from the observer. Refinement will concern only the nearest objects, the most important in terms of QoE.

3.5 Summary and Future Works

This article concerns the problem of progressive 3D mesh transmission. It presents the exploration model, simulation results. The implementation of client-server communication was presented as well as method of progressive data preparation on the server-side and simple communication protocol. The preliminary results of network measurement of 3D mesh progressive transmission in TCP/IP networks was presented and compared to results obtained from simulation studies, conducted previously. Results obtained from the network and simulation evaluations should be taken into consideration when developing the distributed virtual reality applications, using resources distributed across a network (client-server communication) and while working on new network architectures, supporting 3D content storage and transmission.

Further work will include the implementation of the communication between multiple servers and multiple clients as the basis for developing an effective protocol for the efficient development of distributed virtual reality system, based on TCP/IP protocol. The communication protocol will be also extended by transmission of materials and textures.

Acknowledgements. This work was supported with project number N516 482340 "*Experimental station for integration and presentation of 3D views*".

References

1. Nielsen, F.: Visual computing: Geometry, graphics and vision. Charles River Media (2005)
2. Li, H., Li, M., Prabhakaran, B.: Middleware for Streaming 3D Progressive Meshes over Lossy Networks. ACM Transactions on Multimedia Computing, Communications and Applications 2(4), 282–317 (2006)
3. Glomb, P., Nowak, S.: Image coding with contourlet / Wavelet transforms and spiht algorithm:an experimental study. In: Proc. of IMAGAPP 2009, International Conference on Imaging Theory and Applications, Lisboa, Portugal (2009)

4. Skabek, K., Zabik, L.: Implementation of Progressive Meshes for Hierarchical Representation of Cultural Artifacts. Communications in Computer and Information Science (2009)
5. Chen, Z., Barnes, F.J., Bodenheimer, B.: Hybrid and forward error correction transmission techniques for unreliable transport of 3D geometry. Multimedia Systems Journal 10(3) (2005)
6. Al-Regib, G., Altunbask, Y.: 3TP: An application layer protocol for streaming 3-D models. IEEE Transactions on Multimedia 7(6) (2005)
7. Yang, B.L., Li, F., Pan, Z.G., et al.: An Effective Resilent Packetization Scheme for Progressive Mesh Transmission over Unreliable Networks. Journal of Computer Science and Technology 23(6) (2008)
8. Kurose, J.F., Ross, K.W.: Computer Networking: A Top-Down Approach Featuring the Internet. Addison Wesley (2005)
9. Hoppe, H.: Progressive meshes. In: Computer Graphics, SIGGRAPH 1996 Proceedings (1996)
10. OMNet++ homepage, http://www.omnetpp.org
11. INET homepage, http://inet.omnetpp.org
12. Melax, S.: A Simple, Fast and Effective Polygon Reduction Algorithm. Game Developer (11), 44–49 (1998)
13. Skabek, K., Tomaka, A.: Computer Vision for Applications in Medicine and Protection of Monuments. Theoretical and Applied Informatics 22(3) (2010), doi:10.2478/v10179-010-0010-4
14. Nowak, S., Domanska, J., Nowak, M., Glomb, P.: Progressive 3D meshes transmission: traffic generating and simulation evaluation. Theoretical and Applied Informatics 22(4) (2010), doi:10.2478/v10179-010-0012-2
15. Nowak, S., Glomb, P.: Remote Virtual Reality: Experimental Investigation of Progressive 3D Mesh Transmission in IP Networks. In: Proc. HET-NETs (2010)

Chapter 4
Model of the Network Physical Layer for Modern Wireless Systems

Maciej Wrobel and Slawomir Nowak

Abstract. Based on the physical layer model, developed and adopted for discrete event simulators, the preliminary evaluations of the effectiveness of the model, considering the accuracy of the simulation, are presented. We review role of the propagation model, interference model and importance of the mobility of the nodes. We believe that it is important to support decisions of choosing relevant model in discrete events simulations (DES). Such research are important especially in parallel simulations (PDES), because high memory consuming simulations, on the one hand, may lead to high communication overhead and, on the other hand, may be impossible within resources available on one computational node, and therefore may have to be distributed over parallel environment. We assume, that properly developed model (of chosen PHY aspects) in combination with dedicated synchronization method could have strong impact on the performance of the parallel simulation. In the paper we present results of selected efficiency analysis.

4.1 Introduction

Wireless networks are increasingly popular. Highly developed technology of wireless communication not only find applications where wired communication is

Maciej Wrobel · Slawomir Nowak
Academy of Business in Dabrowa Gornicza, Department of Computer Science
ul. Cieplaka 1c, 41-300 Dabrowa Gornicza, Poland
e-mail: wrobelmaciek@gmail.com
http://www.wsb.edu.pl

Slawomir Nowak
Institute of Theoretical and Applied Informatics
Polish Academy of Sciences
Baltycka 5, 44–100 Gliwice, Poland
e-mail: emanuel@iitis.pl

A. Kapczynski et al. (Eds.): Internet - Technical Develop. & Appli. 2, AISC 118, pp. 39–48.
springerlink.com

expensive, but also becomes more popular just for convenience of users. Nowadays wireless networks become important part of business.

Because of popularity of the wireless technologies both industry and academic society need good methods for simulating wireless networks, former for wireless infrastructure planning and later, for research and analysis of the network protocols and technology. High complexity of the wireless technology and high number of clients make simulations very resource-demanding.

Discrete event simulations (DES) has appeared as the most convenient approach for the performance evaluation of network protocols and architectures. Several wireless network DES have been proposed. Examples are NS2[17], GTNetS[7], and some popular extensions of OMNeT++[18]: MobilityFramework[14], MiXiM[15] etc.

To provide useful results simulations should be as realistic as possible. On the other hand the higher realism usually leads to the performance degradation.The popular approach is to simplify the physical layer (PHY) model to reduce the number of events and computational complexity. Simplifying the complexity of the model, however, leads to less accurate simulation but the scale of simulated scenarios increases. Different simulation tools are different in degrees of detail of physical layer implementation and the complexity of the wireless physical layer enforces the use of simplified models, as the tradeoff between the accuracy and the scalability of simulators. The impact of the physical layer modeling accuracy on both the computational cost and the confidence in simulations was investigated and presented in the literature, among others in [8, 10, 12, 13]. Still an open problem is which one of possible PHY models and physical aspects should be selected and implemented for performance reasons ensuring reliability of results for specific applications in simulation.

Because of the potential to achieve a considerable speedup, the parallel simulations (PDES) seem to be particularly interesting[5]. The simulation scenario is divided into a number of logical processes, each of them executes a part of the scenario. However parallel simulation of wireless networks, while maintaining accuracy of the physical layer (PHY) model, is a particularly difficult issue. It is mostly because of the shared type of the medium and the resulting intensity of communication between objects. We assume, that properly developed model (of chosen PHY aspects) have strong impact on the performance of the parallel simulation.

The difficulty of simulations of the wireless systems and formulation of a model is due to the complexity of technologies used in PHY layer of modern wireless systems. Additionally they works on different time scales: in technologies of keying and modulations, such as OFDM[9], SC-FMDA[11] crucial effects happen in time scale of bit (or symbol) transmission (microseconds) and are of few order shorter than frame transmission time (which is of order of milliseconds). In multiple input - multiple output (MIMO) systems time scales are even shorter - the technology is bases on shifts of the signal of order of carrier electromagnetic wave frequency and may be in order of nanoseconds[6]. On the other hand, to perform simulations for analysis of work of higher layers of the networks in reasonable time one have to perform them at least at the time scale of data frame transmissions. To issue that problem one have to use strongly limiting and simplifying assumptions about model, which may have effect of strongly divergent from experiment result.

Present paper is structured as follows. In the next section we present the model, which is a base of our efficiency analysis. In the following part we we present the efficiency analysis of the interference model, the propagation model and role of the mobility in the network. In the last part we present our conclusions and propose further research.

4.2 The Model

In our related paper[20] we proposed the model of the PHY layer of the wireless systems. The model allows modular and flexible implementation of the selected aspects of PHY in the DES (and hopefully in PDES) simulators. To allow analysis of various models it consists of formal equations, which gives possibility to interchange models of physical effects in simulations, without changing other parameters of the simulated system.

To handle with various time scales, needed to analyse modern networks, description of the physical phenomena is based on the linear equations of the amplitudes of electromagnetic wave at given antenna i:

$$R_i(t) = \sum_{k=1}^{\infty} \sum_{j=1}^{M} h_{ijk}(t) T_j(t - \delta t_{ijk}) + n_i(t) \tag{4.1}$$

where $T_j(t')$ is an amplitude of the signal emitted by the j antenna at time $t' < t$, t_{ijk} is a time of signal propagation between the transmitter j and the receiver i through path as k. $h_{ijk}(t)$ coefficient describes path-loss of the amplitude of the signal on the path k. $n_i(t)$ describes inner antena noise and environment-generated noise.

Physical properties of the propagation medium are hidden in the h_{ijk} elements. They factorize to:

$$h_{ijk}(t) = P_0 \cdot p_k(\mathbf{x_i}, \mathbf{x_j}) \cdot v(\mathbf{v_i}, \mathbf{v_j}) \cdot r(\theta_{ij}, \phi_{ij}) \cdot r(\theta_{ji}, \phi_{ji}) \cdot \mathscr{M}, \tag{4.2}$$

where p function describes path loss between receiver i and transmitter j, v is a function that describes effects of mobility, and r describes spatial distribution of the antenna radiation in direction described by angles θ, ϕ and $\mathscr{M}$ describes other phenomena, that influences amplitude of the received signal.

To allow computer analysis of the equation, one may rewrite it in a form:

$$R_i(t) = \sum_{j=1}^{N} h_{ij}(t) T(t - \delta t_{ij}) + n_i(t), \tag{4.3}$$

where N is number of radiation sources (which may be either antennas or scatterers).

To perform simulations at frame transmission timescale, we use relation between amplitude of the electromagnetic wave and power of received signal:

$$P_i \sim |R_i|^2, \tag{4.4}$$

and we introduce phase of the signal ϕ_i, to make at least partially possible analysis of the effects of phase-shift of the electromagnetic wave in either multiple input or multiple output systems (MI/MO).

Depending on the phase of the signals one may reduce number of received signals to signals, that neither interfere constructively, nor destructively. Then, to include various modulation, keying and multi-access techniques in frame-rate timescale we assume that a Bit Error Rate (BER) may be calculated from a function G dependant only on powers of incoming signals which are proportional to h_{ij}:

$$G(t) = G(h_{ij}(t)). \tag{4.5}$$

Explicit form of the $G(t)$ depends on the used modulation, keying and multi-access techniques, and usually is calculated from signal to noise (SNR) or signal to interference and noise (SINR) value.

Our model allows to perform analysis of some quite realistic network models (such as a simplified 802.11a/g network model)[20], but thanks to its formal description, we may change chosen physical effect model to analyse its performance.

4.3 Results

We have implemented presented model in SimPy environment[19]. SimPy is a discrete event simulation framework designed for the Python programming language. Interpreted nature of the Python language allows easy modularisation of the simulation and changeability of physical models. Moreover, SimPy structure allows to implement wireless network simulator without concern on inner elements of simulator such as simulation events and synchronisation messages. Therefore implementation of the model needed little effort comparing to lower level languages, such as C or Fortran.

Modular structure of our simulator, presented on figure 4.1. allows to test various models of physical effects without changes in other parts of the simulation and therefore verify their impact on the simulation efficiency.

Our model allows analysis of the simulation performance with dependancy on the selected model of physical phenomena. In our investigations we performed analysis of:

- role of the propagation model,
- interference model,
- effects of inclusion of mobile nodes in the simulation.

Simulations were performed on personal computer with Intel Core2 Quad Q6600 processor with 2GB of RAM, on Ubuntu 10.04 64bit Linux distribution, and presented time results refer to this architecture.

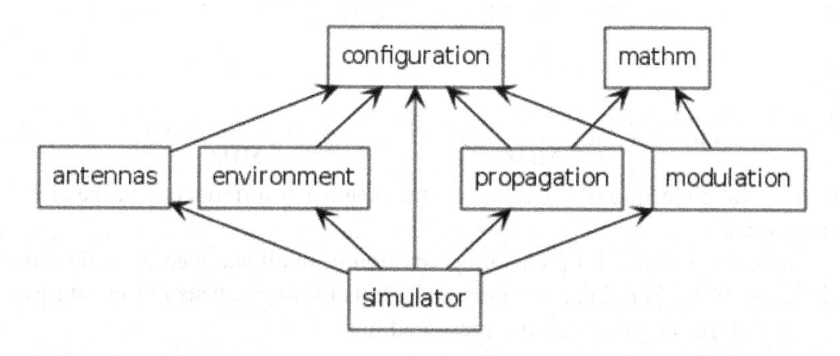

Fig. 4.1 Dependency structure of the modules comprising the implementation. mathm module contains supplementary mathematical methods. Other modules contains description of the physical layer parts corresponding to their names.

4.3.1 Propagation Model

To test impact of the propagation model on the efficiency of the simulation we performed analysis of scenario of communication 10 pairs of transmitter-receiver, all of which are in each other range. Each transmitter had tried to send bunch of 1000 frames to assigned receiver. We tested performance of scenarios with models of propagation given by free-space model[1], radio-engineer model (also known as the Friis transmission equation)[2] and Hata model[4].

Free-space model is a popular description of the path loss, based on physical properties of electromagnetic wave emitted by omnidirectional antenna in an empty space (thus the model neglets effects of reflection, multipath propagation, obstacles etc.). In free-space model path-loss P expressed in dB is proportional to $\log(\frac{d}{\lambda})$:

$$P \sim \log\left(\frac{d}{\lambda}\right), \tag{4.6}$$

where d is distance between transmitter and receiver and λ is a wavelength of the electromagnetic wave.

Radio engineer model is another simple model, in which path-loss measured in dB is proportional to $\log((\frac{d}{\lambda})^2)$:

$$P \sim \log\left(\left(\frac{d}{\lambda}\right)^2\right), \tag{4.7}$$

and is derived with use of the properties of the communicating antenas.

Most complicated analysed model, known as Hata model for ubran areas and is model based on extensive field measurements. Mathematical formulae for dB path-loss is as following:

$$P = 69.55 + 26.16 \log \frac{f}{\text{MHz}} - 13.82 \log \frac{h_T}{m} - a(h_R) + (44.9 - 6.55 \log \frac{h_T}{m}) \log \frac{d}{km}, \quad (4.8)$$

where:

$$a(h_R) = (1.1 \log \frac{f}{\text{MHz}} - 0.7) \frac{h_R}{m} - (1.56 \log \frac{f}{\text{MHz}} - 0.8), \quad (4.9)$$

h_R (H_T) is a height of the receiver (transmitter), and f is a frequency of the electromagnetic wave.

Hata model is suitable for path-loss prediction in small and medium cities and is one of the most used propagation models. Results of our analysis of the simulation efficiency with different models are presented in table 4.1.

Table 4.1 Performance of the propagation models for free-space, Hata and Radio-Engineer (RE) model, measured as system time, needed to evaluate parameters for one simulation event. Time is expressed in seconds. For ease of comparison ratio of evaluation time to the evaluation time of the free-space model is presented.

measure	free-space	Hata	RE
time/event	$4.02 \cdot 10{-}5$	$5.02 \cdot 10^{-5}$	$4 \cdot 10^{-5}$
ratio (to free-space)	1	1.249	0.995

Our results shows that simpler propagation models (i. e. RE, free-space) are more efficient that more complicated ones. While on the first sight difference seems to be important, we found that time needed to evaluate path-loss with given propagation model is almost of one order smaller than time needed to calculate pairwise distances, which is in average $3.4 \cdot 10^{-4}$s.

We believe that the reason lies in vector-based computations of the distance, while path-loss is calculated on scalars, which requires less computation effort. Because of time needed to calculate distances between communicating nodes we found that model of propagation is not very important in the simulation efficiency. We belive that similar results should hold as long as propagation model is based just on distance between nodes. Moreover, similar results are expected for case of statistical shadowing and fading effects inclusion in the model, introduced as multiplicative random factor in path-loss calculation[8], and therefore are computationally simple. On the other hand, effects based on vector-based calculations for each pair of nodes (such as Doppler effect dependant on velocity of the nodes) possibly are more important in efficiency of simulation.

4.3.2 *Interference Model*

The second investigated case was the interference model. During communication each node may receive not only expected message, but also messages addressed to other nodes. In simulations such effect may be included in many ways and one believes, that it is important factor of the simulation efficiency[8], because they may have highest computational complexity of $O(n^2)$. To verify impact of the interference calculations we performed scenarios of simulations with three different models of the interference (similar to those presented in e.g. [8]): scenario with no interference included (thus without any packet collisions, model 1), and two models with packet error rate (PER) calculated from SINR: in one case PER was calculated basing on the maximum value of SINR during packet reception (model 2). In the other case PER was calculated basing on all values of SINR during packet reception, with weight of time of given SINR duration (model 3). In our simulation scenarios nodes are in each other range.

Our results are presented in table 4.2. As one may see times needed to evaluate SINR are shortest for model 1 (and in dedicated simulation may be equal 0), and subsequently increases for model 2 and 3. Higher time of the SINR evaluation for the model 3 than the model 2 raises from that one has to store all values of SINR for given packet in the model 3, while in the model 2 only one value has to be stored. Even bigger difference may be observed between times of PER evaluation for model 2 and 3 – in model 2 (and also in our implementation in model 1) PER is evaluated basing on one value of PER, which is simple scalar calculation, while for model 3 in simulation must be evaluated expression involving whole array of SINR values.

Table 4.2 Times of evaluation interference effects for given scenario for three different models: model 1 (no interference), model 2 (interference based on maximum value of SINR) and model 3 (interference based on all values of SINR). Nonzero times of SINR evaluation in model 1 raises from calls to the dummy procedure realizing SINR, which always return 0. To highlight proportion between evaluation time of various model in the table are presented ratios of the evaluation time to the time of the model 1 evaluation.

calculated part	model 1	model 2	model 3
SINR[s]	0.17	4.14	10.4
SINR ratio	1	24.3	61.2
PER[s]	1.33	1.28	10.41
PER ratio	1	0.96	8.13
serviced messages	107804	60540	107800

The results supports statement of high importance of interference evaluation in case of each-other visibility. Moreover, usage of more complex model causes high degradation of the simulation efficiency. We believe that the model of interference strongly influences simulation accuracy - our preliminary research showed that statistics of communication messages were very similar between model without interference and with full interference (model 1 and model 3), while simplified model of interference (model 2) behaved very different. While this results are still under investigations they shows that decision of choosing interference model may very hardly affects the results and thus simulations should be verified against different interference models.

4.3.3 *Influence of the Mobility*

The mobility of the clients leads to twofold complications in simulation:

- Doppler effect in estimation of the link parameters,
- recalculation all parameters of the links at each step of mobile nodes.

We analysed both effects. Doppler effect bases on vector-based calculations. It requires computation time of same order as distance computation and approximately doubles time required to calculate link parameters. While it seems to impact strongly the simulation efficiency, our investigations showed that the other problem – recalculation of all link parameter – influences simulation efficiency mostly. In the simulation where all nodes are stationary, link parameters do not change during simulations and therefore may be pre-calculated at the beginning of the simulation. Thus execution time of the simulation in most scenarios are dependant on the PHY model only by an additive factor. In the case of the mobile nodes all link parameters may change becase of node movement and in each step they have to be recalculated. Therefore PHY model evaluation gets into simulation execution time as a multiplicative factor, which may have huge impact on the simulation performance (table 4.3).

Table 4.3 Impact of the mobility on the efficiency of the simulation in identical simulation scenario, with only difference based on the recalculation of the link parameters in case of mobile stations in each step of the simulation.

measure	no mobile stations	mobile stations
simulation time (arb. units)	1	17.87
evaluation of the link parameters	0.025	16.49

Second important effect of the mobility may be observed in MI/MO systems. In systems where techniques based on the electromagnetic wave phase are used, movement at distance of order of parts of the wavelength may have important effects on the link quality. Therefore it may be necessary to sample time between simulation events to recalculate link parameters, which may have dramatic effect on simulation performance.

Importance of the mobile stations may be critical especially in parallel simulations. To perform reasonable simulations of the network in parallel environment one has to dispatch simulated nodes over computational clusters[3]. Because of range-related sight range of the stations most of clustering algorithms depends on the geographical positions of the nodes, which change in a mobile environment. Therefore in many scenarios one has to recluster nodes during the simulation and to propagate information about velocity and position of the nodes. In large systems it may cause huge overhead in communication between computational clusters.

4.4 Summary

In present paper we analyse impact of the PHY models on the simulation efficiency. We review role of the propagation model, interference model and importance of the mobility of the nodes. Accuracy has a variable cost depending on the considered PHY aspect. We found that the most important efficiency factor is the node mobility. Causes by far the largest computational overhead, since each node movement causes the necessity recalculation of the link parameters. Moreover, in MIMO systems it may be necessary to sample inter-event time to reevaluate link parameters due to node mobility, which may cause further performance drop. Finally, mobility in parallel discrete event simulations (PDES) may cause high communication overhead due to necessity of propagation all nodes position and velocity over computational clusters and possibly redistribution of the nodes among computational clusters during the simulation.

Our research showed that calculations of the path-loss may vary strongly, depending on the propagation model, but because of the scalar nature of that calculations they are rather small in comparison with the distance and Doppler-effect calculations, which are vector based. Therefore in simulations, where propagation model is based on scalar properties of the nodes it is not very important efficiency factor and in most cases it is better to use realistic models than simple models to obtain more accurate results without large efficiency impact.

Further results of our simulations shows that interference model may have large influence on the efficiency of the simulation. In our investigations we observed that interference model may change dramatically obtained results for the same scenario. Therefore we state that in research of higher network layers with use of DES one has to double-check role of the interference between nodes.

In our research we found that in simulation with high density of the transmitted messages the bottleneck is in sorting of the event queue of the simulator. In some of our simulated scenarios calculations related to PHY model taken much below 10% of the simulation time. In such cases one has to consider using of as accurate PHY models as possible, because their overall impact over simulation time may be negligible.

The described results are consistent with the intuitive assessment of the complexity of the various aspects of the PHY and are in line with other similar studies[8]. Our research is a part of preparation of methods to parallelize simulations (PDES) of modern wireless networks. Efficient parallel simulation of wireless networks, while maintaining accuracy physical layer model, is a particularly difficult issue. We assume, that properly developed model could have strong impact on the performance of the parallel simulation. The developed model will be implemented in a distributed simulator that uses a new, time-stepped method of synchronization[16]. We continue our research to analyse not only computational complexity, but also memory, communication and synchronization demands of various PHY models.

Acknowledgements. The work presented in the paper was sponsored by Polish Ministry of Science and Higher Education grant no. N N516 407138.

References

1. Balanis, C.A.: Antenna Theory. John Wiley & Sons, Nowy Jork (1997)
2. Bakshi, K.A., Bakshi, A.V., Bakshi, U.A.: Antennas And Wave Propagation, Pune, India (2008)
3. Bononi, L., DAngelo, G., Donatiello, L.: HLA-based Adaptive Distributed Simulation of Wireless Mobile Systems. In: PADS 2003: Proceedings of the Seventeenth Workshop on Parallel and Distributed Simulation. IEEE Computer Society, Washington, DC,USA (2003)
4. Digital mobile radio towards future generation systems. COST 231 Final Report (May 10, 2011), http://www.lx.it.pt/cost231/final_report.htm
5. Fujimoto, R.: Parallel and Distributed Simulation Systems. John Wiley and Sons, Inc. (2000)
6. Gesbert, D., et al.: From Theory to Practice: An Overview of MIMO Space=Time Coded Wireless Systems. IEEE Journal on Sel. Areas in Comm. 21(3) (2003)
7. GTNetS homepage, http://www.ece.gatech.edu/research/labs/MANIACS/GTNetS/
8. Ben Hamida, E., et al.: Impact of the Physical Layer Modeling on the accuracy and scalability of Wireless Network Simulation. Transactions of the Society for Modeling and Simulation International (2009)
9. Hanzo, L., et al.: MIMO-OFDM for LTE, Wi-Fi and WiMAX. A John Wiley and Sons, Ltd. (2011)
10. Heidemann, J.: Effects of Detail in Wireless Network Simulation. In: SCS Comm. Netw. and Distrib. Systems Modeling and Sim. Conference, 26.IX (2000)
11. Holma, H., Toskala, A.: LTE for UMTS-OFDMA and SC_FDMA Based Radio Access. John Wiley and Sons, Ltd. (2009)
12. Johnson, D.B.: Validation of Wireless and Mobile Network Models and Simulation. In: DARPA/NIST Network Simulation Validation Workshop (1999)
13. Kotz, D., Newport, C., Elliott, C.: The mistaken axioms of wireless-network research. Dartmouth College Computer Science Techn. Rep. TR2003-467, Dartmouth (2003)
14. Mobility Framework homepage, http://mobility-fw.sourceforge.net/
15. MiXiM homepage, http://mixim.sourceforge.net
16. Nowak, S., Nowak, M., Foremski, P.: New synchronization method for the parallel simulations of wireless networks. In: The 11th International Conference on Next Generation Wired/Wireless Advanced Networking (will apear in LNCS), Sankt Petersburg (2011)
17. The network simulator, ns-2, http://www.isi.edu/nsnam/ns/
18. OMNeT++ homepage, http://inet.omnetpp.org/
19. SimPy homepage, http://www.simpy.sourceforge.net
20. Wrobel, M., Nowak, S.: Model of the network physical layer for modern wireless systems. Theoretical and Applied Informatics 23(2), 147–159 (2011), doi:10.2478/v10179-011-0010-z

Chapter 5
Column-Oriented Metadata Organization of Vision Objects

Malgorzata Bach, Adam Duszenko, and Aleksandra Werner

Abstract. Evolution of an Internet had measurable influence on progress in sphere of vision systems. For 20 years, from the first introduction in the world plug-in camera, this branch of the technology incredible widened one's possibilities. Nowadays vision systems are widely used in many areas of human life. They are applied, for example, to industrial quality assurance, burglar alarm, intelligent control systems of traffic control, car parks monitoring or to the border crossings control. The activity of such systems is concerned with the need of large amounts of LOB data types storage. In order to satisfy efficient access to such data, it's necessary to develop adequate mechanisms of data storage and processing. Because of the fact traditional relational systems may be inefficient, the possibility of using the column-oriented database that may speed up the video content management is proposed. Desired efficiency is achieved by images and their metadata separation, and loading the latter (that is: data about data) directly to column-oriented structures. Thus, the search performance of video stored entirely in the single node of the relational databases and with usage of replicas reside in column-oriented databases, was compared.

5.1 Introduction

As opposed to classic systems of data acquisition and processing (e.g. financial and accounting systems with simple data types) the vision systems operate on data with completely different granularity of information[1]. In conventional systems, the data is simply transformed into information owing to its high granularity that results from a detailed data decomposition just at the stage of its inserting. These data has a simple structure, which is usually a specific value or a string. Systems that operate

Malgorzata Bach · Adam Duszenko · Aleksandra Werner
Silesian University of Technology, Gliwice, Poland
e-mail: Malgorzata.Bach@polsl.pl, Adam.Duszenko@polsl.pl,
Aleksandra.Werner@polsl.pl

[1] Granularity is implied as an amount of data that is required to store the needed information.

A. Kapczynski et al. (Eds.): Internet - Technical Develop. & Appli. 2, AISC 118, pp. 49–55.
springerlink.com

on image data, unlike the classic ones, very often contain data with much lower, than desired, granulation. In this case, extraction of data - often connected with the large amounts of video data management - is time-consuming task, requiring a large amount of computing and storage resources. The problem solution may be an appropriate set of data preparation - so-called metadata, describing selected features of registered objects that causes the search tasks among large number of images will be effective.

5.2 Problem Description

Images metadata, especially semantic metadata, has - like in the classic data models - the high granulation of information and - as a single, specific value - has their unambiguous interpretation. The example can be the city monitoring system entries, where the images of vehicles passing the crossroad are recorded. If - as a result of image recognition - in the video recording metadata, the information about the registered vehicles registration numbers is stored, the sample task "search the specific car within a certain time" will no longer require the on-line analysis of all the images in the collection. After analyzing other, similar to described, examples (industrial processes monitoring, video surveillance of borders, etc.) the metadata numerical growth seemed to help for speeding up the performance of certain types of queries. However, such a statement would be considerable abuse.

Despite the advantageous resulting from the nature of metadata that is represented by the value of simple type (INTEGER or CHARACTER), it is stored with the large multiplicity. It causes that a full description of one image can require the tens or hundreds of additional data generation. Consequently it follows the search space expanding, during the descriptions of objects searching.

The metadata modeling to the simplest structure, allowing its efficient indexing, is inadequate. Theore, other features that will efficiently manage "data about data" and will increase the effectiveness of video surveillance systems by speeding up the search operation, should be looked for.

The images metadata is rather changeless. Generated once for the image, does not change or update and the basic type of performed operation is reading. Hence, the reasonable solution seems to be using column-oriented DBMS. This kind of databases store data tables as sections of columns of data, rather than as rows of data. This has a number of advantages. If a search is being done for items matching a particular value in a column of data, only the storage objects corresponding to this data column within the table, need to be accessed. A traditional row-based database reads the whole table top to bottom.

In the column-based structure, consecutive duplicates within a single column will be automatically removed and null values will not be recorded since the missing record ID implies a null value.

Let's assume, data about different vision devices (ie. cameras) is stored in a database. The physical organization of a data in classic database is shown in Figure 5.1A, while its organization in column-oriented database - in Figure 5.1B.

ID	ANGLE	DISTANCE
1	100	1
2	120	1.2
3	75	0.75
4	95	0.5

A

1, 100, 1;
2, 120, 1.2;
3, 75, 0.75;
4, 95, 0.5;

B

1, 2, 3, 4;
100, 120, 75, 95;
1, 1.2, 0.75, 0.5;

Fig. 5.1 Row and column-oriented database

Suppose, the table T has 100 columns (t1, t2, ..., t100). If data is stored in the traditional way (in rows) and there are a lot of numbers of columns, it may indicate, for example, that only two columns can fit one page. Imagine that number of records equals 1000 and table data is allocated in 500 pages. This means, the sample query SELECT t1, t2 FROM T requires reading 500 pages from a disk (a lot of I/O operations). In the column-oriented database, performing the same query (with assumption, that one column takes 5 pages) will require reading only 10 pages from a disk. In column-oriented database, the number of pages fetched from a disk is much more lower, than in pure relational databases.

5.3 Solution Description

The idea of the proposed solution is to store images and their descriptions (metadata) in a traditional - row-oriented - relational database (in research it was IBM DB2). As far as replicas for metadata are concerned, they are created in a column-oriented databases (Sybase IQ system was used).

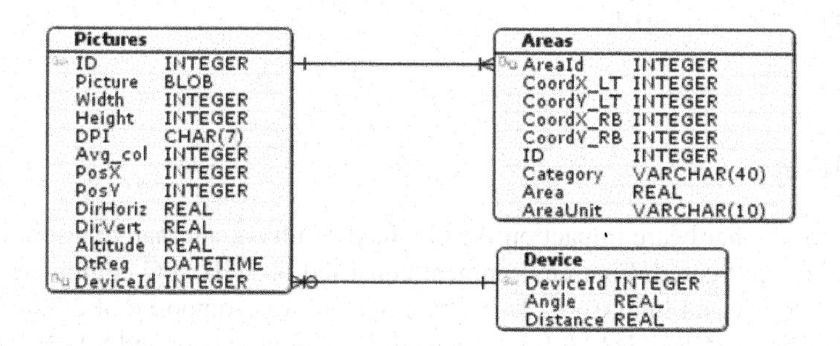

Fig. 5.2 The schema of tested database

It was assumed that the photos will not only be analyzed, but also processed in order to distinguish some interesting for observer objects (AREAS). Therefore, the one picture is related to a lot of areas, as it is shown in Figure 5.2 (1:N relationship). Images are recorded by a camera (table DEVICE). All tables were replicated, but it's worth noting, that in the column-oriented copy of PICTURES table, BLOBs weren't stored.

The attributes of PICTURES table are mostly the photos parameters - e.g. image resolution (DPI), width, height of image, etc, but there are also another data - such as the location of the camera - it means GPS data (PosX, PosY) and extra information about the direction of the device, at the moment of taking the photo (DirHoriz, DirVert). The angle and distance of camera's view are recorded in DEVICE table. The PICTURES table is a type of temporal table - i.e. it's constantly loaded with new "static" (it means: known at the time of data insertion[2] or resulting from the features of vision hardware[3]) data and no rows of data (images) are removed.

For the study it was assumed, the camera is on Polish territory. So, all measured coordinates are within the limits of about 48°-55° (latitude) and about 14°-25° (longitude).

The typical query that searches described data concerns finding the ID of all the shots from the cameras, that recorded the traced (e.g. suspicious) object. In this case, the query sent to the database contains compound conditions, relate (inter alia) to the direction of photo taking (DirHoriz and DirVert attributes) and the coordinates of both - object (attributes: CoordX_LT CoordY_LT, etc.) and camera (PosX, PosY). These values are analyzed taking into consideration the angle of camera's view, in order to determine whether the specific object is within camera shot. The traced object can be found by performing the following SQL query[4]:

```
SELECT ID
FROM PICTURES P, DEVICE D
WHERE P.DEVICEID=D.DEVICE_ID
AND 1<SIN(90-DIRHORIZ-ANGLE/2)*(51-POSX)/(COS(90-DIRHORIZ-ANGLE/2))
AND 50>(51-POSX)*SIN(90-DIRHORIZ+ANGLE/2)/(COS(90-DIRHORIZ+ANGLE/2))
AND
SQRT(DISTANCE)<SQRT(POWER(((51-POSX)*72),2)+POWER(((22-POSY)*111),2))
AND DTREG>'2011-01-01';
```

5.4 Tests

To observe the hardware impact on the obtained results, 2 alternative test environments were configured. First environment consisted of 2GHz PC computers, with 4GB of memory and Intel Core2 Duo CPU. Second was composed of 2.8 GHz PC computers, with 8GB of RAM, Intel Centrino 2 V-Pro processor and SATA 320 GB hard disk.

[2] An example might be the attribute DtReg - the timestamp the photo was taken.

[3] For example DPI or Altitude parameters, where the last one indicates the altitude of a camera at a given moment.

[4] The values 111 and 72 are the lengths (in kilometers) of 1°- respectively - on the meridian and in latitude.

For research purposes, different types of SQL queries were performed, and the number of table's rows varied from 500 000 up to 3 000 000. The size of DEVICE table was negligibly small in comparison with the size of other database tables (remained at the level of a few dozen rows). Thus, the number of table rows, marked on the charts, refers only to AREAS and PICTURES tables. The execution time of sample query (presented in previous section) is shown in Figure 5.3. To better visualize the differences in speed of query execution in IQ and DB2 database systems, the cumulative chart was used.

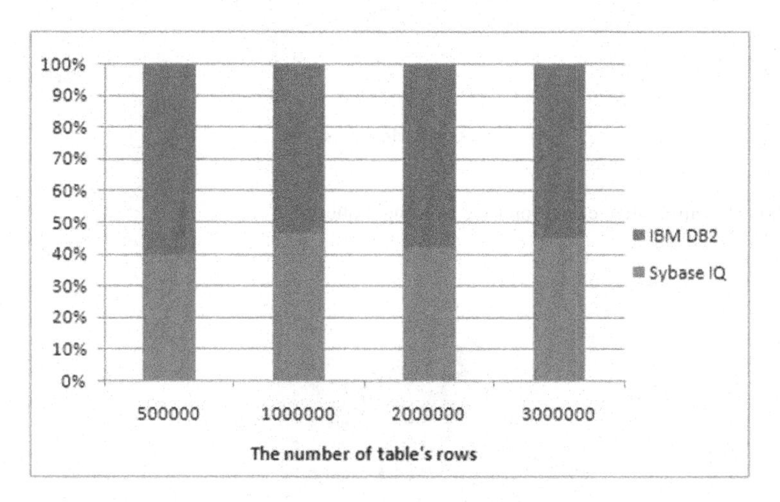

Fig. 5.3 The cumulative execution time of sample query

Other queries performed during the tests, showed the similar speedup characteristics as it was noticed in a previous example. For instance the time executions collected for query:

```
SELECT COUNT(ID)
FROM PICTURES
WHERE 0.2<(51-POSX)*SIN(90-DIRHORIZ/4-35)/(COS(90-DIRHORIZ/4-35))
AND 90>(51-POSX)*SIN(90-DIRHORIZ/4+35)/(COS(90-DIRHORIZ/4+35))
AND 1<((51-POSX)*72)*((51-POSX)*72)+((22-POSY)*111)*((22-POSY)*111);
```

is shown in Figure 5.4.

The performance growth was observed in the variety types of database queries (with/without tables joining, aggregate functions, etc) in every hardware environment, but it significantly depends on the number of rows processed by a query. The most satisfactory speeding up was achieved for the number of 500 000 table's rows (the time of Sybase IQ record searching was about 5,03 times faster than it was in

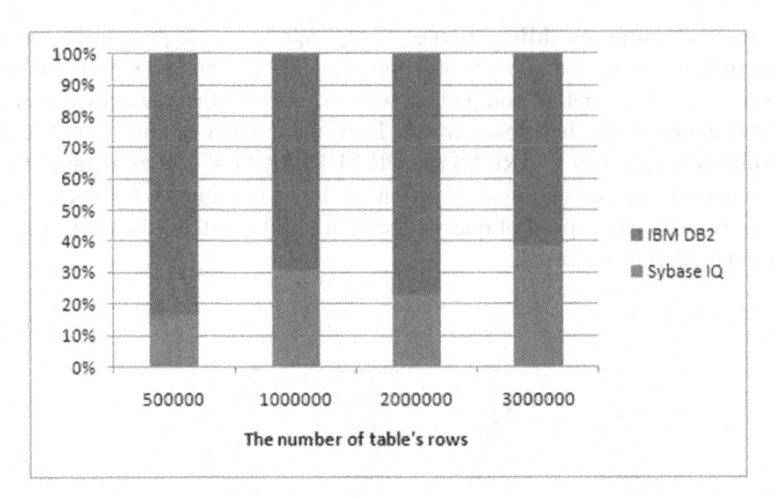

Fig. 5.4 The cumulative execution time of sample query 2

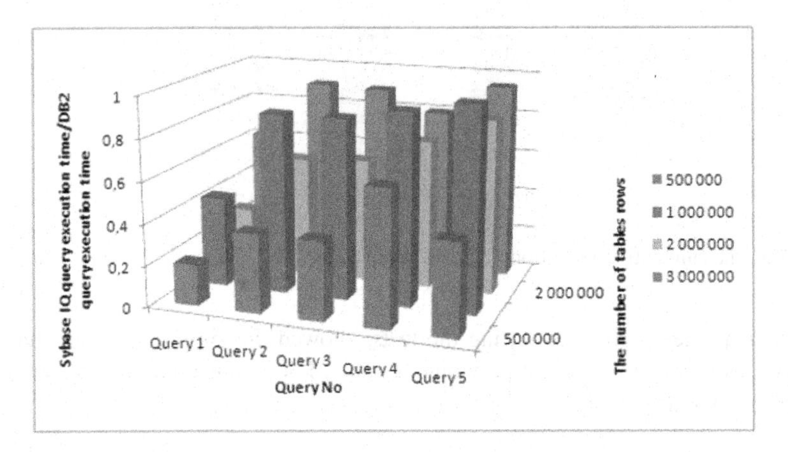

Fig. 5.5 Comparision of the execution time of 5 sample queries

DB2), while the worst - only 9% of speeding up - for the number of 3 000 000 table's rows (Figure 5.5).

It's worth emphasizing, that the Sybase IQ is a purpose-built column-based analytics database. Thus the frequency of data updating is rather low - as opposed to the data insertion and further data search. In order to confirm its pure-analytical purpose, several tests were performed, updating the large amounts of Sybase IQ database data. The differences were already noticeable when updating a single IQ table row. In Sybase IQ the update operation lasted more than 10% slower than it was measured in DB2 database system.

5.5 Summary

The relational representation of multimedia data consists of several attributes, but relatively small subset of them, is used in a typical queries. Due to the numerous rows in the analyzed tables, the influence research of data organization to the performance of its processing seemed to be reasonable. The tests results indicated that column-oriented data storage significantly speeds up the search operations. The performance increase was observed in the variety types of database queries, but speeding up (in some cases) was not satisfactory, so in further tests, in-memory databases and NoSQL databases will be analyzed.

References

1. Chong, R.F., Liu, C., Qi, S.F., Snow, D.R.: Zrozumieć DB2. Nauka na przykładach. PWN (2006)
2. DB2 9.7 Discovery Kit. Wydawnictwo Software Press (2009)
3. DB2 9.7 Software Developers Journal Extra nr 35, Sofware, Warszawa (2009) ISSN: 1734-7661
4. Sybase IQ Administration Guide, http://manuals.sybase.com/onlinebooks/group-iqarc/iqg1121e/iqadmin/GenericBookTextView/63
5. TALOS project, http://talos-border.eu
6. Stonebraker, M., Abadi, D.J., Batkin, A., Chen, X., Cherniack, M., et al.: C-Store: A Columnoriented DBMS. In: VLDB, pp. 553-564 (2005), http://db.csail.mit.edu/projects/cstore/vldb.pdf
7. Vesset, D.: Reorienting the Data Warehouse: Improving the Business Analytics Capabilities with Sybase IQ Columnar Database (2010), IDC, http://www.dnm.ie/documents/whitepapers/Sybase_case_study_Reorienting_the_Data_Warehouse.pdf
8. Imai, Y., Sugiue, Y., Andatsu, A., Masuda, S.: Development of web-based monitoring system with database and homing facilities (2005) ISBN: 972-8924-02-X, IADIS, http://www.iadis.net/dl/final_uploads/200507C025.pdf
9. Gedikli, S., Bandouch, J., Hoyningen-Huene, N., Kirchlechner, B., Beetz, M.: An Adaptive Vision System for Tracking Soccer Players from Variable Camera Settings (2007), http://ias.cs.tum.edu/publications/pdf/gedikli07adaptive.pdf
10. Abuzar, M., Whelan, B., Lyle, G.: Guidelines for Metadata and Data Directory, GRDC SIP 2009 (2009), http://www.usyd.edu.au/agriculture/acpa/documents/metadata_documentation.pdf

Chapter 6
On Controllability of Linear Systems with Jumps in Parameters

Adam Czornik and Aleksander Nawrat

Abstract. In this paper we discuss different types of controllability of jump linear systems. We start with the weakest of the concepts of controllability for jump linear system namely the problem of controllability of the expectation of the final state. The main result of this section gives a necessary and sufficient condition for this type of controllability. Next we consider the possibility of reaching any deterministic target value from given deterministic initial condition in given time with prescribed probability. We have also investigated several variations of this problem such as: the case when the target or initial condition are zero and the case when we want to achieve only certain neighborhood of the target. Part of our results are devoted to the special cases of stochastic controllability. We also consider the concepts of controllability when the time of achieving the target value can be a random variable. Finally we discuss the relationships between the introduced types of controllability.

6.1 Introduction

Modern control systems must meet performance requirements and meantime acceptable behavior even in the presence of abrupt changes in their dynamics due, for instance, to random component failures or repairs, environmental disturbances, dramatic changes in subsystem interconnections, sudden changes in the operating point of nonlinear plant etc. This can be found, for instance, in control of solar systems, robotic manipulator systems, aircraft control systems, large flexible structures for space stations (such as antennas, solar arrays), etc. If abrupt changes have only a small influence on the system behavior, classical sensitivity analysis may provide

Adam Czornik
Silesian University of Technology, Gliwice, Poland
e-mail: Adam.Czornik@polsl.pl

Aleksander Nawrat
WASKO S. A., Berbeckiego 6, 44 - 100 Gliwice, Poland
e-mail: Aleksander.Nawrat@polsl.pl

A. Kapczynski et al. (Eds.): Internet - Technical Develop. & Appli. 2, AISC 118, pp. 57–81.
springerlink.com

an adequate assessment of the effects. When the variations caused by the dynamic changes significantly alter the behavior of the system, a stochastic model that gives a quantitative indication of the relative likelihoods of various possible scenarios would be preferable. In some cases the relevant stochastic model may consist of a linear system with coefficients depending on certain stochastic process. Such models are called jump linear systems. They first appeared in the literature in papers [22] and [15] for continuous time systems. Since then much has been done and a good summary of the results obtained up to 1990 may be found in monograph [23]. Nevertheless still there exists a number of open questions in this area and many results need refinement and improvement. To justify the interest in analysis of this class of control systems let us focus attention on the exemplary situations in which the theory of jump linear systems seems to be the most applicable approach for solving the problem. In order to increase the reliability in the presence of emergency of failures, the control system must provide some kind of fault tolerance. Since von Neuman, we know that redundancy is the basic ingredient in building a reliable systems. The fault-prone control systems have attracted a significant research effort.

To analyze the behavior of such a system we clearly need a description of the occurrence of failures and their influence on the process. Consider, for example, a duplex system where two redundant controllers, C_1 and C_2 are used in parallel to control the plant P. Excluding partial failures of a component, four regimes of operation depending on which controller has failed:

regime 1 C_1 C_2
regime 2 $\underline{C_1}$ C_2
regime 3 C_1 $\underline{C_2}$
regime 4 $\underline{C_1}$ $\underline{C_2}$

Failures thus appear as discrete events that cause a transition, or a jump, of the regime. These events being random, are characterized by transition probabilities $(p_1(n), p_2(n), p_3(n), p_4(n))$, where $p_i\ (n)$ is the probability of regime to be i at moment n. It is reasonable to assume that $p_i\ (n)$ depends on the state of regime at moment $n-1$ and does not depend on states of regimes at moment $k \leq n-2$. In this way we describe regimes as a Markov chain with transition probability

$$P = [p_{ij}]_{i,j=1,\ldots,4},$$

where p_{ij} is the probability that the regime is i under condition that the previous regime was j. A classical model in reliability theory is given as

$$P = \begin{bmatrix} e^{-2\lambda\Delta} & e^{-\lambda\Delta} - e^{-2\lambda\Delta} & e^{-\lambda\Delta} - e^{-2\lambda\Delta} & 1 - 2e^{-\lambda\Delta} + e^{-2\lambda\Delta} \\ 0 & e^{-1} & 0 & 1 - e^{-\lambda\Delta} \\ 0 & 0 & e^{-\lambda\Delta} & 1 - e^{-\lambda\Delta} \\ 0 & 0 & 0 & 1 \end{bmatrix}$$

where $\lambda\Delta$ is the individual failure rate of C_1 and C_2 at time interval Δ. In writing P a simultaneous failure of C_1 and C_2 was excluded (the transition from 1 to 4 has zero probability) which corresponds to the realistic situation where failures are rare events and the occurrence of a simultaneous failure is highly unlikely. The failures manifest themselves through a modification of the actuator-process-sensors cascade. Typically, the failure of a sensor introduces a bias or a drift in one of the measurement variables. Similarly, a failed actuator might produce a constant action on the process regardless of the command signal it receives. A physical fault in some parts of the plant also modifies the dynamics.

Yet another example of this kind has been discussed in [8]. This example deals with reliable control system design. The control system is described by

$$x(k+1) = Ax(k) + B(r(k))\,u(k),$$

where

$$A = \begin{bmatrix} 2{,}71828 & 0 \\ 0 & 0.36788 \end{bmatrix}, \; B(1) = \begin{bmatrix} 1.71828 & 1.71828 \\ -0.63212 & 0.63212 \end{bmatrix}$$

$$B(2) = \begin{bmatrix} 0 & 1.71828 \\ 0 & 0.63212 \end{bmatrix}, \; B(3) = \begin{bmatrix} 1.71828 & 0 \\ -0.63212 & 0 \end{bmatrix}$$

$$B(4) = \begin{bmatrix} 0 & 0 \\ 0 & 0 \end{bmatrix}.$$

This model captures the failure/repair events for a reliable system with two actuators, in which actuators may fail and need to be repaired. State 1 of $r(k)$ represents the case that both actuators work well, states 2 and 3 represent the case where one of actuators fails and has to be repaired, and state 4 represents the case where both actuators fail. Let p_f and p_r denotes the failure rate and repair rate, where the actuators repair and failure events are independent, then the probability transition matrix of Markov chain $r(k)$ is given by

$$P = \begin{bmatrix} (1-p_f)^2 & (1-p_f)\,p_r & (1-p_f)\,p_r & p_r^2 \\ (1-p_f) & (1-p_f)(1-p_r)\;p_r & & p_r(1-p_r) \\ (1-p_f)\,p_f\;p_r p_f & & (1-p_f)\,1-p_r & p_r 1-p_r \\ p_f^2 & (1-p_r)\,p_f & (1-p_r)\,p_f & (1-p_r)^2 \end{bmatrix}.$$

It is thus seen that fault-tolerant control systems are naturally described in terms of Cartesian product of a discrete random jump variable accounting for the occurrence of failures, and the more usual variables in continuous space representing the plant dynamics, i.e. in the terms of system with jumps in parameters.

Another example of this kind was discussed in [25]. Consider a linear system

$$\dot{x}(t) = Ax(t) + Bu(t)$$

with

$$A = \begin{bmatrix} \frac{1}{6} & \frac{1}{5} \\ -\frac{1}{5} & -\frac{1}{6} \end{bmatrix}, \; B = \begin{bmatrix} 1 & 0 \\ 0 & 1 \end{bmatrix}. \tag{6.1}$$

The control which minimizes the quadratic cost

$$\int_0^{\infty} (x'(t)Qx(t) + u'(t)Ru(t))\, dt$$

with

$$Q = \begin{bmatrix} \frac{2}{3} & 0 \\ 0 & \frac{4}{3} \end{bmatrix}, R = \begin{bmatrix} 1 & 0 \\ 0 & 1 \end{bmatrix}$$

is given by

$$u(t) = -Kx(t),$$

where

$$K = \begin{bmatrix} 1 & 0 \\ 0 & 1 \end{bmatrix}.$$

The closed loop system is asymptotically stable. In real system the state vector x is not available. Instead of x we have y -the sensor observation of x which is related to x through the relation

$$y(t) = Lx(t),$$

where

$$L = \begin{bmatrix} l_1 & 0 \\ 0 & l_2 \end{bmatrix}.$$

The system (6.1) remains stable for control

$$u(t) = -Ky(t)$$

if $l_1 > \frac{1}{2}, l_2 > \frac{1}{2}$. Assume that because of sensor failure there is no feedback from x_1 i.e $l_1 = 0$, and the system becomes unstable. Taking into account the possibility of failure of the sensors one can propose as a model the following system

$$\dot{x}(t) = Ax(t) + B(r(t))u(t),$$

where the random variable $r(t)$ takes one of four values 1, 2, 3, 4 and takes value 1 if both sensors work properly, 2 if the first sensor is broken, 3 if the second sensor is broken and 4 if both sensors are broken. The matrices $B(1), B(2), B(3), B(4)$ are as follows

$$B(1) = B,\ B(2) = B \begin{bmatrix} 0 & 0 \\ 0 & l_2 \end{bmatrix}$$

$$B(3) = B \begin{bmatrix} l_1 & 0 \\ 0 & 0 \end{bmatrix},\ B(4) = 0.$$

Again we are led to a hybrid model with continuous dynamics variables perturbed by random transitions of a regime variable which is discrete.

Further discussion of this kind of applications may be found in the following references [26], [27], [7], [28], [29].

In [30] an application of hybrid models of the similar structure has been proposed to control a solar thermal receiver. In this paper a solar 10 MWe electrical generating system has been described. This system has been build in California desert. A field of movable mirrors is used to focus the sun's energy on a central boiler. One of the most important control loops in the boiler is the steam temperature regulator which controls the feedwater flow rate to maintain the proper outlet steam temperature. The steam temperature regulator has been designed on the basis of a linear dynamic model which evolving from both analytical and empirical studies of the dynamic behavior of a solar-powered central receiver. A difficult problem is posed by the motion of clouds over the heliostats. On a partly cloudy day, the clouds tend to cover the helistats for a time that is quite short when compared to the dominant system time constants. These sudden changes in isolation may be frequent and are essentially unpredictable. The relevant system model, depends on the isolation level, thus changes in a discrete and apparently random fashion. Another anomaly is created by the cloud action. The perturbation variables in the system refer to a set of nominal operating points of the nonlinear equation of the system motion. The system variables are continuous across discontinouities in insolation, while the insolation level changes discretely.

In recent years concern with the safety of air traffic near crowded airports and tragic accidents have prompted research into more sophisticated radar tracking algorithms which could help air controllers monitor incoming and outgoing aircrafts. A specific difficulty appears where there exists a heavy traffic of small highly maneuverable private aircrafts interfering with commercial jet liners.

In a military context, a related problem is the so-called evasive target tracking problem where it is desired to keep track of an object which is maneuvering quickly in an attempt to evade its pursuer. The performance of the tracking system heavily depends on the accuracy and sophistication of the model used to describe the target dynamics.

The first element of the model is based on motion equations, relating variables like horizontal position, heading speed, bank angle or flight path angle. Physic provides a set of differential equations in R^n, where n is the number of variables we retain.

A well-known problem when multiple targets are to be tracked is associating of radar data to the various tracks, which might be difficult when a clutter of signals in a given region exists. However, another less usual phenomenon is of primary interest here, namely the effects of sudden changes in the acceleration of the target. The trajectory can be divided into several sequences when the aircraft flies with (almost) constant acceleration, bank angle and flight path angle; regimes of flight typically considered are ascending flight after take-off, turning, accelerated flight, uniform cruise motion or descending approach flight. The transitions between these regimes are discrete and depend primarily on the pilot decision which are in turn influenced by weather data, mission controller indications and on-board information such as fuel consumption, etc. or perception of the threat. Depending on the current regime of flight chosen by the pilot, the coefficients of the dynamic model have to be adjusted.

Of course evasive maneuvers are chosen to confuse the pursuer, and are therefore characterized by frequent, large, irregular and seemingly irrational acceleration changes. From the tracker point of view, these transitions are perceived as random, and a model which makes the least use possible of other a priori information is a hybrid stochastic model with continuous dynamics perturbed by random transitions of a regime variable.

Further discussion of this problem is given in [3]-[5].

Other important applications of jump systems are presented for example in the following papers: [24] (placement of failure-prone actuators on large space structures), [1], [6], [9] (control of manufacturing systems), [31] (analysis of transient electrical power networks), [18], (economic policy planing). Athans in [2] suggested that this model setting also has the potential to become a basic framework in posing and solving control-related issues in Battle Management Command, Control, and Communications (BM/C^3) systems.

Most of the published results deal with continuous-time systems. This is natural because the process variables are continuous. However modern digital applications require discrete-time models. The purpose of this work is to present a comprehensive treatment of mathematical aspects of controllability, stability, and linear quadratic problem for discrete-time jump linear systems. In fact various topics for continuous and discrete-time systems are covered in parallel. It is often believed that results for continuous-time systems are also valid for discrete time systems, but it is not always true. Discrete-time systems have their own specification, and there exists a number of crucial and profound differences between the continuous and discrete-time systems. Controllability belongs to this class of problems. The differences will be transparent to the reader throughout the work.

The work is organized as follows: In paragraph 1 we introduce the system and establish basic notation. Several concepts of controllability of jump linear systems and the relations between them are discussed in paragraph 2. In paragraph 3 the problems of stability and stabilizability are considered.

The research presented here were done as a part of project no. 420/BO/A and was supported by The National Centre for Research and Development funds in the years 2010 - 2012.

6.2 Discrete-Time Jump Linear Systems

By discrete-time jump linear system we understand the following system:

$$x(k+1) = A(r(k))x(k) + B(r(k))u(k), \tag{6.2}$$

where $x(k) \in R^n$ denotes the process state vector, $k = 0, 1, ..., u(k) \in R^m$ is the control input, $r(k)$ is a Markov chain on a probability space $(\Omega, \mathscr{F}, P)$ which takes values in a finite set $S = \{1, 2, ..., s\}$ with transition probability matrix $P = [p(i, j)]_{i,j \in S}$ and initial distribution $\pi = [p(i)]_{i \in S}$. Furthermore, for $r(k) = i$, $A_i := A(i)$ and $B_i := B(i)$ are constant matrices of appropriate sizes. Denote by $x(k, x_0, \pi, u)$ the solution of (6.2) under the control u, with initial condition x_0 at time $k = 0$ and initial distribution of the Markov chain π. In the case when the initial distribution is Dirac it takes the form

$$p(i_0) = 1 \text{ and } p(i) = 0 \text{ for } i \neq i_0 \tag{6.3}$$

for some $i_0 \in S$, we will denote the solution of (6.2) by $x(k,x_0,i_0,u)$. In case of no control ($B(i) = 0$, $i \in S$) the appropriate solutions will be denoted by $x(k,x_0,\pi)$ and $x(k,x_0,i_0)$, respectively. We will consider only deterministic initial condition x_0. The control $u = (u(0),u(1),...)$ is assumed to be such that $u(k)$ is measurable with respect to the $\sigma-$field generated by $r(0)$, $r(1)$, ..., $r(k)$. That is, the control is causal. The assumption of causality of the control may be expressed alternatively by stating that the control $u(k)$ is of the form $f_k(r(0),r(1),...,r(k))$. It is worth to notice that even if the control is of the form $u(k) = f_k(r(k))$ the solution $x(k,x_0,\pi,u)$ is not in general a Markov chain, however the joint process $(x(k,x_0,\pi,u),r(k))$, which takes values in $R^n \times S$, is a Markov chain for any control of the form $u(k) = f_k(r(k))$.

In this formulation the state of the system (6.2) is hybrid. It consists of two parts: x which is continuous and r which is discrete. For example, in target tracking problem x contains target location variables (position and velocity), and the discrete variable r represents presence of a manoeuvre, target classification (friend or foe), etc. In this example u will represent the tracker platform orientation command.

By I_n we denote the identity matrix of size n. For a square matrix A we denote by $\rho(A)$ the spectral radius. For a random variable X and a $\sigma-$field $\mathscr{F}_0$ we denote by EX the expectation and by $E(X|\mathscr{F}_0)$ the conditional expectation. When the $\sigma-$field $\mathscr{F}_0$ is generated by certain random variable Y we write $E(X|Y)$. Moreover for conditional expectation of the form $E(X|r(0))$, where $r(0)$ has distribution π, we introduce symbol $E_\pi X$ and in special case when π is a Dirac distribution of the form (6.3) we write $E_{i_0}X$. We use similar convention for conditional probability denoted by $P(\cdot|A)$, when $A = \{r(0) = i_0\}$ we write $P_{i_0}(\cdot)$.

We introduce also the following notation which is used in formulation of controllability results in chapter 2.

$$F(k,k) = I_{n\times n}$$

$$F(k,l,i_{k-1},...,i_l) = A(i_{k-1})A(i_{k-2})...A(i_l),$$

for all $k > l \geq 0$, $i_{k-1},...,i_l \in S$

$$F(k,l) = A(r(k-1))A(r(k-2))...A(r(l)),\ k > l \geq 0,$$

$$\overline{F}_\pi(k,l) = E_\pi(F(k,l)|r(l-1)), k > l \geq 1$$

$$\overline{F}_\pi(k,0) = E_\pi F(k,0)$$

$$W_\pi(k) = E_\pi \sum_{t=0}^{k-1} \overline{F}_\pi(k,t+1)B(r(t))B'(r(t))\overline{F}'_\pi(k,t+1) \tag{6.4}$$

Using this notation we can write the solution of (6.2) in the following form

$$x(k,x_0,\pi,u) = F(k,0)x_0 + \sum_{t=0}^{k-1} F(k,t+1)B(r(t))u(t), \tag{6.5}$$

or

$$x(k,x_0,\pi,u) = F(k,0,r(k-1),...,r(0))x_0 + \sum_{t=0}^{k-1} F(k,t+1,r(k-1),...,r(t+1))B(r(t))u(t), k \geq 1. \quad (6.6)$$

We also use the following notation

$$\overline{S}_{\pi}^{(N)} = \{i_0,...,i_{N-1} \in S : p(i_0)p(i_0,i_1)...p(i_{N-2},i_{N-1}) > 0\} \quad (6.7)$$

$\overline{\overline{s}}_{\pi}^{(N)}$ is the number of elements of $\overline{S}_{\pi}^{(N)}$, and in the case when the initial distribution π is of the form $p(i_0) = 1$, and $p(i) = 0$ for $i \neq i_0$ we will write $\overline{S}_{i_0}^{(N)}$, and $\overline{\overline{s}}_{i_0}^{(N)}$ instead of $\overline{S}_{\pi}^{(N)}$, and $\overline{\overline{s}}_{\pi}^{(N)}$, respectively. It will be convenient to have the elements of $\overline{S}_{i_0}^{(N)}$ ordered in a sequence. In that purpose let order the elements of the set

$$S^N = \{(i_0,...,i_{N-1}) : i_0,...,i_{N-1} \in S\}$$

as follows

$$(i_0,1,1,...,1,1),(i_0,1,1,...,1,2),...,(i_0,1,1,...,1,s),...$$

$$(i_0,1,1,...,2,1),(i_0,1,1,...,2,2),...,(i_0,1,1,...,2,s),...$$

$$(i_0,s,s,...,s,1),(i_0,s,s,...,s,2),...,(i_0,s,s,...,s,s).$$

Withdraw all the elements $(i_0,...,i_{N-1})$ such that

$$p(i_0,i_1)...p(i_{N-2},i_{N-1}) = 0.$$

The sequence of all elements of $\overline{S}_{i_0}^{(N)}$ obtained in this way is called the natural order in the set $\overline{S}_{i_0}^{(N)}$ and it is denoted by $\widetilde{S}_{i_0}^{(N)}$. Fix a number $N > 0$ and a sequence $(i_0,i_1,...,i_{N-1})$ of elements of S. Consider a matrix column blocks which are numbered successively by sequences: $i_0,\widetilde{S}_{i_0}^{(2)},...,\widetilde{S}_{i_0}^{(N)}$ and the block $(i_0,i_1,...,i_k)$, $k = 0,1,..,N-1$ is given by

$$F(N,k,i_{N-1},...,i_k)B_{i_k}$$

and the others are equal to 0. Denote the matrix obtained in this way by $C(i_0,i_1,...,i_{N-1})$ and by $G(i_0)$ -the matrix consisting of all $C(i_0,i_1,...,i_{N-1})$ (as row blocks numbered by $\widetilde{S}_{i_0}^{(N)}$) for $(i_0,i_1,...,i_{N-1}) \in \overline{S}_{i_0}^{(N)}$. Moreover by $H(i_0) \in R^{n\overline{\overline{s}}_{i_0}^{(N)} \times m}$ let denote a matrix row blocks of which are numbered by the sequence $\widetilde{S}_{\pi}^{(N)}$, the block $(i_0,i_1,...,i_{N-1})$, is given by $F(N,0,i_{N-1},...,i_0)$. For example in the case when $S = \{1,2\}, N = 3$ and $p(i,j) > 0$ for all $i,j \in S$ and $p(1) = 1$, we have

$$G(1) = \begin{bmatrix} C(1,1,1) \\ C(1,1,2) \\ C(1,2,1) \\ C(1,2,2) \end{bmatrix} =$$

$$\begin{matrix} (1) & (1,1) & (1,2) & (1,1,1) & (1,1,2) & (1,2,1) & (1,2,2) \\ [A_1^2B_1 & A_1B_1 & 0 & B_1 & 0 & 0 & 0] \\ [A_2A_1B_1 & A_2B_1 & 0 & 0 & B_2 & 0 & 0] \\ [A_1A_2B_1 & 0 & A_1B_2 & 0 & 0 & B_1 & 0] \\ [A_2^2B_1 & 0 & A_2B_2 & 0 & 0 & 0 & B_2] \end{matrix}$$

and

$$H(1) = \begin{bmatrix} A_1^3 \\ A_2A_1^2 \\ A_1A_2A_1 \\ A_2^2A_1 \end{bmatrix}.$$

Moreover let $f_1^{(k)},...,f_n^{(k)} \in R^{nk}$ denote the vectors defined by

$$f_l = \left. \begin{bmatrix} e_l \\ e_l \\ \vdots \\ e_l \end{bmatrix} \right\} k\text{-times } e_l, \; l = 1,...,n$$

where $e_1,...,e_n$ is the standard base in R^n. Let us denote

$$\mathscr{A}_{i_0}^{(\delta)} = \left\{ X \subset \overline{S}_{i_0}^{(N)} : P_{i_0}\left((i_0, r(1),...,r(N-1)) \in X\right) \geq \delta \right\}$$

and by $\overline{\overline{x}}$ denote the number of elements of $X \in \mathscr{A}_{i_0}^{(\delta)}$. For each $X \in \mathscr{A}_{i_0}^{(\delta)}$ let $G_X(i_0)$ be the submatrix of $G(i_0)$ that consists of the blocks $C(\alpha)$ for $\alpha \in X$ and for $\beta \in \overline{S}_{i_0}^{(N)}$

$$A_\beta = \{\omega \in \Omega : (r(0), r(1),...,r(N-1)) = \beta\}. \tag{6.8}$$

and

$$\delta_0(i_0) = \min\left\{ P_{i_0}(A_\beta) : \beta \in \overline{S}_{i_0}^{(N)} \right\}. \tag{6.9}$$

With each $\alpha \in \overline{S}_\pi^{(N)}$, $\alpha = (i_0,...,i_{N-1})$ we associate a deterministic time-varying system

$$x(k+1) = \overline{A}(k)x(k) + \overline{B}(k)u(k), \; N-1 \geq k \geq 0, \tag{6.10}$$

where

$$\left(\overline{A}(k), \overline{B}(k)\right) = \left(A(i_k), B(i_k)\right), \; N-1 \geq k \geq 0,$$

and we call it a deterministic system which corresponds to α.

6.3 Controllability

Since the early work on state-space approaches to control systems analysis, it was recognized that certain nondegeneracy assumption were useful, in particular in the context of optimal control. However, it was until Kalman's work [19] that the property of controllability was isolated as of interest in and of itself, as it characterizes the degrees of freedom available when attempting to control a system.

The study of controllability for linear systems has spanned a great number of research directions, and topics such as testing degrees of controllability, and their numerical analysis aspects, are still a subject of intensive research.

The idea of controllability of jump linear systems has been already discussed in the following papers [10], [11], [12], [16], [21], in discrete time case and in [13], [23], [17], in continuous time case. The origin of idea of controllability discussed in the papers [21], [23], [17], namely the idea of $\varepsilon-$controllability with probability δ comes from papers [14], [20], [32] where general stochastic systems have been considered. In [16] the authors discussed the original idea of reaching given target in random time, and under different assumptions on the time they obtain different types of controllability. In fact they appear to be equivalent (see section 2.4). The problem of reaching target in the fixed time is disused in [10], [11], where in the first paper the target is a given value of expectation of the state and in the second one a vector. In this chapter we propose certain new ideas of controllability and we make a comparison with the existing ones.

This chapter is organized as follows. In Section 2.1 we study the weakest of the concepts of controllability for jump linear system namely the problem of controllability of the expectation of the final state (controllability with respect to the expectation). The main result of this section, Theorem 6.1 gives a necessary and sufficient condition for this type of controllability. Most of the results are from [10]. In the next section our attention is focussed on the possibility of reaching any deterministic target value from given deterministic initial condition in given time with prescribed probability (stochastic controllability with probability δ). We have also investigated several variations of this problem such as: the case when the target or initial condition are zero (so called controllability to zero or from zero), and the case when we want to achieve only certain neighborhood of the target. Section 2.3 is devoted to the special case of stochastic controllability with probability δ namely to the case $\delta = 1$. Most of the analysis tools presented here were taken from [11]. Whereas in Sections 2.1, 2.2 and 2.3 the control horizon is fixed, in Section 2.4 we consider the concepts of controllability when the time of achieving the target value can be a random variable. The results of this section are partially published in [12]. Finally in Section 2.5 the relationships between the introduced types of controllability are explained as well as the comparison with existing results is made.

6.4 Controllability with Respect to Expectation

In this chapter we propose a definition of controllability indicating the possibility of reachability of any given value of the expectation of the final state in given time.

We start from the following theorem:

Theorem 6.1. *For a fixed initial distribution π of the Markov chain the following conditions are equivalent*

1. *Matrix $W_\pi(N)$ is invertible (definition of $W_\pi(N)$ is given by (6.4)).*
2. *For all $x_1 \in R^n$ there exists a control u such that*

$$E_\pi x(N,0,\pi,u) = x_1. \tag{6.11}$$

3. *For all $x_0, x_1 \in R^n$ there exists a control u such that*

$$E_\pi x(N,x_0,\pi,u) = x_1. \tag{6.12}$$

Definition 6.1. If one of the conditions 1-3 of Theorem 6.1 is satisfied then we call the system (6.2) $\pi-$controllable with respect to the expectation at time N ($\pi-$CWRE at time N). If the system (6.2) is $\pi-$CWRE at time N for each initial distribution then we say that it is controllable with respect to the expectation at time N (CWRE at time N).

Theorem 6.1 gives the necessary and sufficient conditions for $\pi-$CWRE at time N, moreover the proof is constructive in the sense that the control which governs the expectation to the desired value is explicitly given but the disadvantage is that the condition is difficult to check. A sufficient condition easier to check but non constructive in the sense that it does not give the control is presented now.

Corollary 6.1. *If there exists a sequence $(i_0,...,i_{N-1}) \in \overline{S}_\pi^{(N)}$ such that*

$$rank \begin{bmatrix} B_{i_{N-1}} & A_{i_{N-1}}B_{i_{N-2}} \cdots \prod_{j=1}^{N-1} A_{i_j}B_{i_0} \end{bmatrix} = n \tag{6.13}$$

then the system is $\pi-$CWRE at time N. ($A_i = A(i)$, $B_i = B(i)$).

The next example shows that deterministic controllability of each pair (A_i, B_i) is not a necessary condition for system (6.2) to be CWRE at time N.

Example 6.1. Consider system (6.2) with a two-state form structure. Let $p(1) = p(2) = 0.5$, $p(1,1) = p(2,2) = p(2,1) = p(1,2) = 0.5$, and

$$A_1 = \begin{bmatrix} 0 & 2 \\ 1 & 1 \end{bmatrix}, A_2 = \begin{bmatrix} 2 & 1 \\ 1 & 0 \end{bmatrix}, B_1 = \begin{bmatrix} 0 \\ 1 \end{bmatrix}, B_2 = \begin{bmatrix} 0 \\ 0 \end{bmatrix}.$$

If we put $i_0 = i_1 = ... = i_{N-1} = 1$ for $N \geq 2$, then we have

$$p(i_0)p(i_0,i_1)p(i_1,i_2)...p(i_{N-2},i_{N-1}) = (0.5)^N$$

and

$$rank\left[B_{i_{N-1}}\left|A_{i_{N-1}}B_{i_{N-2}}\cdots\prod_{j=1}^{N-1}A_{i_j}B_{i_0}\right.\right]=2.$$

By Corollary 6.1, we conclude that the system is CWRE at time N for each $N \geq 2$.

Theorem 6.2. *The following conditions are equivalent*

1. *Matrix* $W_\pi(N)$ *is invertible (definition of* $W_\pi(N)$ *is given by (6.4)) for each initial distribution* π.
2. *For all* $x_1 \in R^n$ *and all initial distribution* π *there exists a control sequence* $u(k)$, $k = 0, ..., N-1$ *such that*

$$E_\pi x(N, 0, \pi, u) = x_1. \tag{6.14}$$

3. *For all* x_0, $x_1 \in R^n$ *and all initial distributions* π *there exists a control sequence* $u(k)$, $k = 0, ..., N-1$ *such that*

$$E_\pi x(N, x_0, \pi, u) = x_1. \tag{6.15}$$

4. *For all* x_0, $x_1 \in R^n$ *and all* $i_0 \in S$ *there exists a control sequence* $u(k)$, $k = 0, ..., N-1$ *such that*

$$E_{r(0)=i_0} x(N, x_0, i_0, u) = x_1. \tag{6.16}$$

5. *For all* $x_1 \in R^n$ *and all* $i_0 \in S$ *there exists a control sequence* $u(k)$, $k = 0, ..., N-1$ *such that*

$$E_{r(0)=i_0} x(N, 0, i_0, u) = x_1. \tag{6.17}$$

6. *Matrix* $W_{r(0)=i_0}(N)$ *is invertible for all* $i_0 \in S$.

The next examples show that deterministic controllability of each pair (A_i, B_i) is not a sufficient condition for $\pi-$CWRE at time N. We have already shown (Example 6.1) that deterministic controllability of each pair (A_i, B_i) is not a necessary condition for $\pi-$CWRE at time N. However, Corollary 6.1 guarantees that controllability of at least one deterministic system corresponding to $\alpha \in \overline{S}_\pi^{(N)}$ is a sufficient condition for $\pi-$CWRE at time N but it is not a necessary condition. It is possible (see, Example 6.3) that the deterministic system that corresponds to each $\alpha \in \overline{S}_\pi^{(N)}$ is not controllable in the deterministic sense and the system is $\pi-$CWRE at time N.

Example 6.2. Consider system (6.2) with a two-state form structure. Let $p(1) = 1$, $p(2) = 0$, $p(1,1) = p(2,2) = 0$, $p(2,1) = p(1,2) = 1$, and

$$A_1 = \begin{bmatrix} 0 & 1 \\ 0 & 0 \end{bmatrix}, A_2 = \begin{bmatrix} 0 & 0 \\ 1 & 0 \end{bmatrix}, B_1 = \begin{bmatrix} 0 \\ 1 \end{bmatrix}, B_2 = \begin{bmatrix} 1 \\ 0 \end{bmatrix}.$$

To test $\pi-$CWRE at time N observe that $\overline{F}(N,l)$ is in this case a constant random variable and

$$E\overline{F}(N,l)B(r(l))B'(r(l)) =$$

$$\overline{F}(N,l)E\left(B(r(l))B'(r(l))\right) = \begin{bmatrix} 0 & 0 \\ 0 & 0 \end{bmatrix}.$$

Consequently $W_\pi(N) = 0$ and the system is not $\pi-$CWRE at time N for each N though each pair (A_i, B_i), $i = 1,2$ is controllable, and consequently it is not CWRE at time N for each N.

Example 6.3. Consider system (6.2) with a two-state form structure. Let $p(1) = 1$, $p(2) = 0$, $p(1,1) = p(2,2) = p(2,1) = p(1,2) = 0.5$, and

$$A_1 = \begin{bmatrix} 1 & 0 \\ 0 & 1 \end{bmatrix}, A_2 = \begin{bmatrix} 0 & 0 \\ 0 & 0 \end{bmatrix}, B_1 = \begin{bmatrix} 0 \\ 1 \end{bmatrix}, B_2 = \begin{bmatrix} 1 \\ 1 \end{bmatrix}.$$

Then $\overline{S}_\pi^{(2)} = \{(1,1),(1,2)\}$ and neither the deterministic system that corresponds to the element $(1,1)$ nor the one which corresponds to the element $(1,2)$ is controllable in the deterministic sense. However,

$$W_\pi(2) = \begin{bmatrix} 1 & 1 \\ 1 & 3 \end{bmatrix} > 0$$

and this system is $\pi-$CWRE at time 2.

6.5 Stochastic Controllability

In this paragraph we examine a concept of controllability idea of which is to steer any deterministic initial condition to a given deterministic target value at given time with prescribed probability. We have the following definition:

Definition 6.2. We say that system (6.2) is stochastically controllable with probability δ at time N (SCWP δ at time N) if, for all $x_0, x_1 \in R^n$ there exists a control u such that

$$P_\pi(x(N, x_0, \pi, u) = x_1) \geq \delta. \tag{6.18}$$

Analogically, we say that system (6.2) is SCWP δ at time N to zero (from zero) if, for all $x_0 \in R^n$ ($x_1 \in R^n$) there exists a control u such that

$$P_\pi(x(N, x_0, \pi, u) = 0) \geq \delta \qquad (P_\pi(x(N, 0, \pi, u) = 0) \geq \delta) \tag{6.19}$$

In the case when $\delta = 1$ we say that system (6.2) is directly controllable (DC) at time N (DC at time N to zero, DC at time N from zero).

The next theorem reduces problems of SCWP δ at time N and DC at time N for system (6.2) with initial distribution of the Markov chain π, to problems of SCWP δ at time N and DC at time N for system (6.2) with Dirac initial distribution.

Theorem 6.3. *Suppose that for each $i \in S$ system (6.2) with $P(r(0) = i) = 1$ is SCWP $\delta(i)$ at time N (SCWP $\delta(i)$ at time N to zero, SCWP $\delta(i)$ at time N from* *zero) then for initial distribution π of the form $P(r(0)=i)=p(i), i\in S$ it is SCWP δ at time N (SCWP δ at time N to zero, SCWP δ at time N from zero), where*

$$\delta = \sum_{i\in S} p(i)\delta(i).$$

Moreover system (6.2) is DC at time N (DC at time N to zero, DC time N from zero) for all initial distributions π if and only if it is DC at time N (DC at time N to zero, DC time N from zero) for all Dirac initial distributions.

Having in mind the previous result we may restrict our considerations to system (6.2) with initial distribution of the Markov chain being a Dirac one, without losing generality and in the remainder of the section it is assumed that the distribution is of the form: $P(r(0)=i_0)=1$.

The next theorem contains necessary and sufficient conditions for SCWP δ at time N as well as SCWP δ at time N from zero and to zero.

Theorem 6.4. *System (6.2) is SCWP δ at time N from zero if and only if there exists $X\in\mathscr{A}_{i_0}^{(\delta)}$ such that*

$$rankG_X(i_0) = rank\begin{bmatrix} G_X(i_0) & f_l^{(\bar{\bar{x}})} \end{bmatrix}, \text{ for all } l=1,...,n. \tag{6.20}$$

System (6.2) is SCWP δ at time N to zero if and only if there exists $X\in\mathscr{A}_{i_0}^{(\delta)}$ such that

$$ImH_X(i_0) \subset ImG_X(i_0), \tag{6.21}$$

and it is SCWP δ at time N if and only if there exists $X\in\mathscr{A}_{i_0}^{(\delta)}$ such that

$$rankG_X(i_0) = rank\begin{bmatrix} G_X(i_0) & f_l^{(\bar{\bar{x}})} \end{bmatrix}, \text{ for all } l=1,...,n, \tag{6.22}$$

and

$$ImH_X(i_0) \subset ImG_X(i_0). \tag{6.23}$$

Remark 6.1. Suppose that the system is SCWP δ at time N and let $X\in\mathscr{A}_{i_0}^{(\delta)}$ be such that (6.22) and (6.23) hold. It easy to conclude that for each $\alpha\in X$ the deterministic system that corresponds to α is controllable. In fact conditions (6.22) and (6.23) are much stronger. They imply that if we fix $x_0,x_1\in R^n$, and $\alpha,\beta\in X$ of the following form $\alpha=(i_0,i_1,...,i_{N-1}), \beta=(j_0,j_1,...,j_{N-1}), i_l=j_l$ for $l=0,...,k$ then it is possible to construct the controls $u_\alpha=(u_\alpha(0),...,u_\alpha(N-1))$, $u_\beta=(u_\beta(0),...,u_\beta(N-1))$ which steer x_0 and x_1 in deterministic systems corresponding to α and β, respectively and the controls are such that $u_\alpha(l)=u_\beta(l)$ for $l=0,...,k$.

In the previous papers devoted to $\varepsilon-$controllability with probability δ the definition is formulated as follows (see [14], [20], [21], [32]):

For all $x_0, x_1 \in R^n$ there exists a control u such that

$$P\left(\|x(N,x_0,i_0,u)-x_1\| < \varepsilon\right) \geq \delta.$$

The first impression is that this definition is less restrictive than SCWP δ at time N. The following theorem establishes the equivalence between this definition for $x_0 = 0$ and SCWP δ at time N from zero.

Theorem 6.5. *Fix $\varepsilon > 0$. If for all $x_1 \in R^n$ there exists a control u such that $P\left(\|x(N,0,i_0,u)-x_1\| < \varepsilon\right) \geq \delta$ then system (6.2) is SCWP δ at time N from zero.*

Remark 6.2. In a very similar way we can show that if for all $x_0, x_1 \in R^n$ ($x_0 \in R^n$) there exists a control u such that

$$P\left(\|x(N,x_0,i_0,u)-x_1\| < \varepsilon\right) \geq \delta$$

($P\left(\|x(N,x_0,i_0,u)\| < \varepsilon\right) \geq \delta,$) then system (6.2) is SCWP δ at time N, (SCWP δ at time N to zero).

6.5.1 Direct Controllability

The case $\delta = 1$ of SCWP δ at time N deserves special attention. Therefore we have introduced the notion DC for this type of controllability and now we will focus our attention on it.

Since $\overline{S}_{i_0}^{(N)}$ is a finite set, there exists a number $\delta_0 > 0$ such that

$$\overline{S}_{i_0}^{(N)} = \mathscr{A}_{i_0}^{(\delta)} \text{ for all } \delta \geq \delta_0. \tag{6.24}$$

Having that in mind we obtain necessary and sufficient conditions for DC at time N from Theorem 6.4. They are given in the following Corollary.

Corollary 6.2. *System (6.2) is DC to zero at time N if and only if*

$$ImH_{\overline{S}_{i_0}^{(N)}}(i_0) \subset ImG_{\overline{S}_{i_0}^{(N)}}(i_0) \tag{6.25}$$

System (6.2) is DC from zero at time N if and only if

$$rankG_{\overline{S}_{i_0}^{(N)}}(i_0) = rank\begin{bmatrix} G_{\overline{S}_{i_0}^{(N)}}(i_0) & f_l^{\left(\overline{s}_{i_0}^{(N)}\right)} \end{bmatrix}, \tag{6.26}$$

for all l=1,...,n. System (6.2) is DC at time N if and only if

$$rankG_{\overline{S}_{i_0}^{(N)}}(i_0) = rank\begin{bmatrix} G_{\overline{S}_{i_0}^{(N)}}(i_0) & f_l^{\left(\overline{s}_{i_0}^{(N)}\right)} \end{bmatrix}, \tag{6.27}$$

and

$$ImH_{\overline{S}_{i_0}^{(N)}}(i_0) \subset ImG_{\overline{S}_{i_0}^{(N)}}(i_0) \tag{6.28}$$

for all l=1,...,n.

From this theorem it is clear that (6.2) is DC at time N if and only if it is simultaneously DC to zero and from zero.

When we consider the system without jumps it is well known that controllability from zero implies the controllability to zero and that inverse implication is not true. The next example shows that for the system with jumps the π–direct controllability from zero does not imply the π–direct controllability to zero.

Example 6.4. Consider the system (6.2) with $S=\{1,2\}, N=2$,

$$A_1=\begin{bmatrix}1 & 2\\ 3 & 1\end{bmatrix}, A_2=\begin{bmatrix}-1 & 2\\ 1 & -1\end{bmatrix}, B_1=\begin{bmatrix}0\\ 2\end{bmatrix}, B_2=\begin{bmatrix}0\\ 1\end{bmatrix}$$

$$p(i)>0,\ p(i,j)>0,\ i,\ j\in S.$$

According to the notation we have (for simplicity we omit the index $\overline{S}_{i_0}^{(2)}$ in $G(i)$ and $H(i)$)

$$G(1)=\begin{bmatrix}C(1,1)\\ C(1,2)\end{bmatrix}=\begin{bmatrix}A_1B_1 & B_1 & 0\\ A_2B_1 & 0 & B_2\end{bmatrix}=\begin{bmatrix}4 & 0 & 0\\ 2 & 2 & 0\\ 4 & 0 & 0\\ -1 & 0 & 1\end{bmatrix}$$

and

$$G(2)=\begin{bmatrix}C(2,1)\\ C(2,2)\end{bmatrix}=\begin{bmatrix}A_1B_2 & B_1 & 0\\ A_2B_2 & 0 & B_2\end{bmatrix}=\begin{bmatrix}2 & 0 & 0\\ 1 & 2 & 0\\ 2 & 0 & 0\\ -1 & 0 & 1\end{bmatrix}$$

and it is easy to check that condition (6.26) is satisfied. The control which steers zero initial condition to $\begin{bmatrix}x_1^{(0)}\\ x_2^{(0)}\end{bmatrix}$ at time $N=2$ is given by

$$u(0)=\begin{cases}\frac{x_1^{(0)}}{4} & \text{if} \quad r(0)=1\\ \frac{x_1^{(0)}}{2} & \text{if} \quad r(0)=2\end{cases}$$

$$u(1)=\begin{cases}\frac{x_2^{(0)}}{2}-\frac{x_1^{(0)}}{4} & \text{if} \quad r(0)=1, r(1)=1\\ x_2^{(0)}+\frac{x_1^{(0)}}{4} & \text{if} \quad r(0)=1, r(1)=2\\ x_2^{(0)}+\frac{x_1^{(0)}}{2} & \text{if} \quad r(0)=2, r(1)=2\\ \frac{x_2^{(0)}}{2}-\frac{x_1^{(0)}}{4} & \text{if} \quad r(0)=2, r(1)=1\end{cases}.$$

From the other hand the system is not π–DC to zero at time 2. In fact we have

$$H(1) = \begin{bmatrix} 7 & 4 \\ 6 & 7 \\ 5 & 0 \\ -2 & 1 \end{bmatrix}, H(2) = \begin{bmatrix} 3 & -4 \\ -2 & 3 \\ 1 & 0 \\ -2 & 5 \end{bmatrix}$$

and

$$\begin{bmatrix} 11 \\ 13 \\ 5 \\ 1 \end{bmatrix} \in H(1), \begin{bmatrix} -1 \\ 1 \\ 1 \\ 3 \end{bmatrix} \in H(2)$$

but

$$\begin{bmatrix} 11 \\ 13 \\ 5 \\ 1 \end{bmatrix} \notin G(1), \begin{bmatrix} -1 \\ 1 \\ 1 \\ 3 \end{bmatrix} \notin G(2).$$

Example 6.5. Consider the system (6.2) with $S = \{1,2\}, N = 3$,

$$A_1 = \begin{bmatrix} 0 & 1 \\ 1 & 0 \end{bmatrix}, A_2 = \begin{bmatrix} 1 & 0 \\ 0 & 1 \end{bmatrix}, B_1 = \begin{bmatrix} 0 \\ 1 \end{bmatrix}, B_2 = \begin{bmatrix} 0 \\ 0 \end{bmatrix},$$

$$p\ [p(i,j)]_{i,j=1,2} = \begin{bmatrix} 0.3 & 0.7 \\ 0 & 1 \end{bmatrix}.$$

If the initial distribution is of the form $\pi : p(2) = 1$, then the problem is trivial because $\overline{S}_2^{(3)} = \{(2,2,2)\}$. Consider the case when $p(1) = 1$. We have

$$\overline{S}_1^{(3)} = \{(1,1,1),(1,1,2),(1,2,2)\}$$

and

$$P_1\left(A_{(1,1,1)}\right) = 0.09,\ P_1\left(A_{(1,1,2)}\right) = 0.21,\ P_1\left(A_{(1,2,2)}\right) = 0.7.$$

Moreover for $0.09 \geq \delta > 0$

$$\mathscr{A}_1^{(\delta)} = \left\{\overline{S}_1^{(3)}, \{(1,1,2),(1,2,2)\}, \{(1,1,1),(1,2,2)\}, \{(1,2,2)\}, \right.$$
$$\left. \{(1,1,1),(1,1,2)\}, \{(1,1,2)\}, \{(1,1,1)\}\right\},$$

for $0.21 \geq \delta > 0.09$

$$\mathscr{A}_1^{(\delta)} = \left\{\overline{S}_1^{(3)}, \{(1,1,2),(1,2,2)\}, \{(1,1,1),(1,2,2)\}, \{(1,2,2)\}, \right.$$
$$\left. \{(1,1,1),(1,1,2)\}, \{(1,1,2)\}\right\},$$

for $0.3 \geq \delta > 0.21$

$$\mathscr{A}_1^{(\delta)} = \left\{\overline{S}_1^{(3)}, \{(1,1,2),(1,2,2)\}, \{(1,1,1),(1,2,2)\}, \{(1,2,2)\}, \right.$$

$$\{(1,1,1),(1,1,2)\}\},$$

for $0.7 \geq \delta > 0.3$

$$\mathscr{A}_1^{(\delta)} = \left\{\overline{S}_1^{(3)}, \{(1,1,2),(1,2,2)\}, \{(1,1,1),(1,2,2)\}, \{(1,2,2)\},\right.$$

for $0.79 \geq \delta > 0.7$

$$\mathscr{A}_1^{(\delta)} = \left\{\overline{S}_1^{(3)}, \{(1,1,2),(1,2,2)\}, \{(1,1,1),(1,2,2)\}\right\}$$

for $0.91 \geq \delta > 0.79$

$$\mathscr{A}_1^{(\delta)} = \left\{\overline{S}_1^{(3)}, \{(1,1,2),(1,2,2)\}\right\}$$

for $1 \geq \delta > 0.91$

$$\mathscr{A}_1^{(\delta)} = \left\{\overline{S}_1^{(3)}\right\}.$$

$$C(1,1,1) = \begin{bmatrix} 0\,1\,0\,0\,0\,0 \\ 1\,0\,0\,1\,0\,0 \end{bmatrix},$$

$$C(1,1,2) = \begin{bmatrix} 1\,0\,0\,0\,0\,0 \\ 0\,1\,0\,0\,0\,0 \end{bmatrix}$$

$$C(1,2,2) = \begin{bmatrix} 0\,0\,0\,0\,0\,0 \\ 1\,0\,0\,0\,0\,0 \end{bmatrix}$$

$$G(1) = \begin{bmatrix} 0\,1\,0\,0\,0\,0 \\ 1\,0\,0\,1\,0\,0 \\ 1\,0\,0\,0\,0\,0 \\ 0\,1\,0\,0\,0\,0 \\ 0\,0\,0\,0\,0\,0 \\ 1\,0\,0\,0\,0\,0 \end{bmatrix},$$

$$H(1) = \begin{bmatrix} 0\,1 \\ 1\,0 \\ 1\,0 \\ 0\,1 \\ 0\,1 \\ 1\,0 \end{bmatrix},$$

Let us first discuss the problem of *DC* at time 3. From the structure of the matrix $C(1,2,2)$ we see that conditions (6.20) and (6.21) can not be satisfied with X such that $(1,2,2) \in X$ and therefore the system is not *DC* at time 3. Moreover it is not SCWP δ at time 3 for all $\delta > 0.7$, because for each such δ and $X \in \mathscr{A}_1^{(\delta)}$ we have $(1,2,2) \in X$. Now for $0.7 \leq \delta < 0.79$ we see that conditions (6.20) and (6.21) could be satisfied only with $X = \{(1,1,1),(1,1,2)\}$. However, we have

$$G_X(1) = \begin{bmatrix} 0\,1\,0\,0\,0\,0 \\ 1\,0\,0\,1\,0\,0 \\ 1\,0\,0\,0\,0\,0 \\ 0\,1\,0\,0\,0\,0 \end{bmatrix} \text{ with } rank G_X(1) = 3$$

and

$$rank \left[G_X(1)\ f_1^{(\bar{x})} \right] = 4.$$

So the system is not SCWP δ at time 3 from zero for $\delta = 0.7$. Moreover we have

$$Im H_X(1) = Im \begin{bmatrix} 0\,1 \\ 1\,0 \\ 1\,0 \\ 0\,1 \end{bmatrix} \subset Im G_X(1)$$

therefore the system is SCWP δ at time 3 to zero for $\delta = 0.7$. Finally notice that conditions (6.20) and (6.21) are satisfied with $X = \{(1,1,2)\}$ and therefore the system is SCWP δ at time 3 for all $\delta < 0.79$, because for such δ we have $\{(1,1,2)\} \in \mathscr{A}_1^{(\delta)}$.

6.6 Controllability at Random Time

In this paragraph we will discuss the problem of controllability at random time. The idea of the next definition is taken from [16].

Definition 6.3. The system (6.2) is $\pi-$weakly controllable, if for all x_0, $x_1 \in R^n$ there exists a control u and a random time τ a.s. finite such that

$$P_\pi\left(x(\tau, x_0, \pi, u) = x_1\right) > 0; \tag{6.29}$$

$\pi-$controllable, if this probability can be made equal to one; $\pi-$strongly controllable if it is weakly $\pi-$controllable and $ET_{x_0,x_1} < \infty$, for each x_0, $x_1 \in R^n$ where

$$T_{x_0,x_1} = \min\left\{k : x(k, x_0, \pi, u) = x_1\right\}. \tag{6.30}$$

Analogically, we introduce the concepts of π-weak controllability, π- controllability and π-strong controllability to zero and from zero. As usually in the case of Dirac initial distribution π of the form $p(i_0) = 1$ we will say about i_0-weak controllability, i_0-controllability and i_0-strong controllability. If system (6.2) is π-weakly controllable, π-controllable or π-strongly controllable (to zero, from zero) for all initial distribution π we will call it weakly controllable, controllable or strongly controllable (to zero, from zero), respectively.

Remark 6.3. From the definition it is clear that $\pi-$controllability (to zero, from zero) implies $\pi-$ weak controllability (to zero, from zero). It is also true that $\pi-$strong controllability implies $\pi-$controllability. To prove this statement suppose that (6.2) is $\pi-$strong controllable. Fix x_0, $x_1 \in R^n$ and let control u be such that $ET_{x_0,x_1} < \infty$ with T_{x_0,x_1} given by (6.30). Suppose that

$$P_\pi\left(x\left(T_{x_0,x_1},x_0,\pi,u\right)=x_1\right)<1.$$

Then

$$P_\pi(A)>0, \tag{6.31}$$

where

$$A=\left\{\omega\in\Omega : x\left(T_{x_0,x_1},x_0,\pi,u\right)\neq x_1\right\}.$$

Moreover for $\omega\in\Omega$ we have

$$T_{x_0,x_1}=\infty. \tag{6.32}$$

(6.31) together with (6.32) implies that $ET_{x_0,x_1}=\infty$. This contradicts the assumption about $\pi-$strong controllability. Of course the same is true for $\pi-$strong controllability to zero (from zero) and $\pi-$controllability to zero (from zero).

The next theorems show that the problem of $\pi-$weak controllability (to zero, from zero), can be reduced to the problem of i_0-weak (to zero, from zero) for i_0 such that $p(i_0)>0$.

Theorem 6.6. *System (6.2) is $\pi-$weakly controllable (to zero, from zero) if and only if there exists $i_0\in S_\pi^{(1)}$, such that (6.2) is i_0-weakly controllable (to zero, from zero).*

The next two theorems establish the relationships between $\pi-$ controllability (to zero, from zero), $\pi-$ strong controllability (to zero, from zero) and $i-$ controllability (to zero, from zero), $i-$ strong controllability (to zero, from zero).

Theorem 6.7. *System (6.2) is $\pi-$ controllable (to zero, from zero) if and only if for all $i\in S_\pi^{(1)}$ system (6.2) is $i-$controllable (to zero, from zero).*

Theorem 6.8. *System (6.2) is $\pi-$strongly controllable (to zero, from zero) if and only if it is $i-$strongly controllable (to zero, from zero) for all $i\in S_\pi^{(1)}$.*

Having in mind the previous three theorems and the Definition 6.3 we can formulate the following remark.

Remark 6.4. System (6.2) is weakly controllable (controllable, strongly controllable) if and only if it is $i-$weakly controllable ($i-$controllable, $i-$strongly controllable) for all $i\in S$. The same is true for controllability of all these types to zero and from zero.

The next theorem shows that to check weak controllability or weak controllability to zero it is enough to check $i-$weak controllability or $i-$weak controllability to zero for recurrent state $i\in S$.

Theorem 6.9. *System (6.2) is weakly controllable (to zero) if only if for each recurrent state i (6.2) is $i-$weakly controllable (to zero)*

Notice that the proof does not work for controllability from zero. The reason is that even if we take $x_0=0$ we cannot guarantee that $x\left(\tau_{i_0},x_0,i_0,0\right)=0$.

Our next goal is to show that weak controllability, controllability and strong controllability are equivalent. We will show the equivalence in the following way: first we will show a necessary condition for weak controllability (Theorem 6.10) and next we will prove that this condition is sufficient for strong controllability (Theorem 6.11). According to Remark 6.3 this means that the three concepts are equivalent.

Theorem 6.10. *The necessary condition for weak controllability and weak controllability from zero of the system (6.2) is that for all closed communicating classes C of S there exists a sequence* $(i_0,...,i_{T-1}) \in C^T$ *such that*

$$p(i_0,i_1)...p(i_{T-2},i_{T-1}) > 0$$

and

$$rank\, L = n,$$

where

$$L = \left[F(T,T)B_{i_{T-1}}, F(T,T-1,i_{T-1})B_{i_{T-2}}, F(T,1,i_{T-1},...,i_1)B_{i_0}\right].$$

Theorem 6.11. *If for all closed communicating classes C of S there exists a sequence* $(i_0,...,i_{T-1}) \in C$ *such that* $p(i_0,i_1)...p(i_{T-2},i_{T-1}) > 0$ *and*

$$rank\, L = n \tag{6.33}$$

where

$$L = \left[F(T,T)B_{i_{T-1}},\ F(T,T-1,i_{T-1})B_{i_{T-2}}, F(T,1,i_{T-1},...,i_1)B_{i_0}\right]$$

then the system (6.2) is strongly controllable.

As we have mentioned before from Theorem 6.10 and Theorem 6.11 the following theorem follows:

Theorem 6.12. *For system (6.2) the weak controllability, controllability and strong controllability are equivalent. Moreover each of these conditions is equivalent to the following: for all closed communicating classes C of S there exists a sequence* $(i_0,...,i_{T-1}) \in C$ *such that* $p(i_0,i_1)...p(i_{T-2},i_{T-1}) > 0$ *and*

$$rank\, L = n. \tag{6.34}$$

where

$$L = \left[F(T,T)B_{i_{T-1}}, F(T,T-1,i_{T-1})B_{i_{T-2}}, F(T,1,i_{T-1},...,i_1)B_{i_0}\right].$$

Remark 6.5. From Theorem 6.10 we know that condition (6.34) is a necessary condition for weak controllability from zero and from Theorem 6.11 that it is a sufficient condition for strong controllability, and in particular for strong controllability from zero. Therefore weak controllability from zero, controllability from zero and strong controllability from zero are equivalent, and they are equivalent to (6.34) too.

Remark 6.6. Having in mind the interpretation of condition (6.34) for deterministic time-varying system that corresponds to the sequence

$$(i_0,...,i_{T-1})$$

(see, (6.2)) we can reformulate the condition for weak controllability, controllability and strong controllability by saying that for all closed communicating classes C of S there exists a sequence

$$(i_0,...,i_{T-1}) \in C$$

such that

$$p(i_0,i_1)...p(i_{T-2},i_{T-1}) > 0$$

and the deterministic time varying system that corresponds to

$$(i_0,...,i_{T-1})$$

is controllable in the deterministic sense.

We can use the same arguments as in the proofs of Theorem 6.10 and Theorem 6.11 to show that for system (6.2) weak controllability to zero, controllability to zero and strong controllability to zero are equivalent and each of them is equivalent to the following: for all closed communicating classes C of S there exists a sequence $(i_0,...,i_{T-1}) \in C$ such that $p(i_0,i_1)...p(i_{T-2},i_{T-1}) > 0$ and the deterministic time varying system that corresponds to $(i_0,...,i_{T-1})$ is controllable to zero in deterministic sense. Therefore we have the following result.

Theorem 6.13. *For system (6.2) the weak controllability to zero, controllability to zero and strong controllability to zero are equivalent. Moreover each of these conditions is equivalent to the following: for all closed communicating classes C of S there exists a sequence $(i_0,...,i_{T-1}) \in C$ such that $p(i_0,i_1)...p(i_{T-2},i_{T-1}) > 0$ and*

$$ImF(T,0,i_{T-1},...,i_0) \subset Im\left[F(T,T)B_{i_{T-1}}...F(T,1,i_{T-1},...,i_1B_{i_0}\right].$$

It appears that the properties of i-strong controllability and i-strong controllability to zero are properties of class of state. More precisely we have the following theorem.

Theorem 6.14. *If C is a closed commutating class of state and system (6.2) is i-strongly controllable (i-strongly controllable to zero) for certain $i \in C$ then it is j-strongly controllable (j-strongly controllable to zero) for all $j \in C$.*

The equivalence enables us to consider in the remaining part only the concept of strong controllability or strong controllability to zero.

To demonstrate our considerations we present an example.

Example 6.6. Consider the system (6.2) with $S = \{1,2,3,4,5\}$,

$$A_1 = \begin{bmatrix} 0 & 1 \\ 0 & 0 \end{bmatrix}, A_2 = \begin{bmatrix} 0 & 1 \\ 0 & 0 \end{bmatrix}, A_3 = \begin{bmatrix} 0 & 1 \\ 1 & 0 \end{bmatrix},$$

$$A_4 = \begin{bmatrix} 0 & 0 \\ 0 & 1 \end{bmatrix}, A_5 = \begin{bmatrix} 0 & 0 \\ 1 & 1 \end{bmatrix}$$

$$B_1 = \begin{bmatrix} 0 \\ 1 \end{bmatrix}, B_2 = \begin{bmatrix} 0 \\ 0 \end{bmatrix}, B_3 = \begin{bmatrix} 1 \\ 0 \end{bmatrix}, B_4 = \begin{bmatrix} 1 \\ 1 \end{bmatrix}, B_5 = \begin{bmatrix} 1 \\ 1 \end{bmatrix}$$

$$[p(i,j)]_{i,j=1,2,3,4} = \begin{bmatrix} p(1,1) & p(1,2) & p(1,3) & 0 & 0 \\ 0 & 1 & 0 & 0 & 0 \\ 0 & 0 & p(3,3) & p(3,4) & p(3,5) \\ 0 & 0 & p(4,3) & p(4,4) & p(4,5) \\ 0 & 0 & 0 & p(5,4) & p(5,5) \end{bmatrix}.$$

All elements not denoted by 0 are nonzero entries. The state space of this chain consists of two closed communicating classes of recurrent states $\{2\}$ and $\{3,4,5\}$. The state 1 is a transient state.

The system is not strongly controllable. It follows from Theorem 6.12. If we take $C = \{2\}$then there is a unique sequence $(i_0,...,i_{T-1}) \in C^T$ such that $p(i_0,i_1)...p(i_{T-2},i_{T-1}) > 0$, namely $(2,...,2)$ and by Caley-Hamilton Theorem

$$rank\,[F(T,T)B_2 \quad \dots \quad F(T,1,2,...,2)B_2] = rank\,[B_2 \quad A_2B_2],$$

but

$$rank\,[B_2 \quad A_2B_2] = 1.$$

For the same reason the system is not 2-strongly controllable. According to Theorem 6.14 problems of i-strong controllability for $i = 3,4,5$ are equivalent. Using this theorem for sequence $(3,3) \in S_3^{(2)}$ we have

$$rank\,[B_3 \quad A_3B_3] = 2$$

and the system is i-strongly controllable for $i = 3,4,$ and 5.

Now let us discuss the problem of strong controllability to zero. We use theorem Theorem 6.13 with $(2,2) \in S_2^{(2)}$ and obtain

$$ImF(2,0,i_1 = 1,...,i_0 = 1) = ImA_2^2 = \{0\}.$$

Moreover because $i-$strong controllability implies $i-$strong controllability to zero we know that the system is $i-$strongly controllable to zero for $i = 2,3,4,5$ and by Theorem 6.13 it is strongly controllable to zero.

6.7 Comparison and Discussion

Assume that the initial distribution π is given as $p(i_0) = 1$ and try to discuss the relationships between $\pi-$CWRE at time N, SCWP δ at time N and $\pi-$strong controllability. First notice that if the system is SCWP δ at time N for each recurrent i_0 then according to Remark 6.1 there exists $\alpha \in \overline{S}_{i_0}^{(N)}$ such that the deterministic system which corresponds to α is controllable. Therefore by Corollary 6.1 the system is CWRE at time N and strongly controllable. Example 6.3 shows that CWRE at time N does not imply strong controllability. It also demonstrates that neither strong controllability nor CWRE at time N implies DC at time N. It is so because the necessary condition for DC at time N is that all deterministic systems corresponding to each element of $\overline{S}_{i_0}^{(N)}$ are controllable and we have systems without this property which are CWRE at time N (Example 6.3). Moreover, when we consider a Markov chain with one class of transient states then the existence of one element in $\overline{S}_{i_0}^{(N)}$,such that the corresponding deterministic system is controllable, is enough for the strong controllability. Finally, when we have a system which is strongly controllable then there exist natural N and $\delta > 0$ such that it is SCWP δ at time N and, according to the above considerations, CWRE at time N. In fact for fixed x_1, $x_2 \in R^n$ from the strong controllability we conclude that there are a control u and an almost sure finite random variable T such that

$$P\left(x\left(T, x_0, i_0, u\right) = x_1\right) = 1. \tag{6.35}$$

Since T is a.s. finite there exists at least one natural number N such that

$$P\left(T = N\right) := \delta > 0. \tag{6.36}$$

From (6.35) and (6.36) we conclude that the system is SCWP δ at time N.

References

1. Akella, R., Kumar, P.R.: Optimal control of production rate in failure prone manufacturing systems. IEEE Transactions on Automatic Control 31, 116–126 (1986)
2. Athans, M.: Command and control (C2) theory: a challange to control science. IEEE Transactions on Automatic Control 32(4), 286–293 (1987)
3. Bar Shalom, Y., Fortman, T.E.: Tracking and Data Association. Academic Press, New York (1988)
4. Bar Shalom, Y.: Tracking methods in a multitarget environment. IEEE Transactions on Automatic Control 23, 618–628 (1978)
5. Bar Shalom, Y., Birmiwal, K.: Variable dimension for maneuvering target tracking. IEEE Transactions on Aero. Electr. Systems, AES 18, 621–628 (1982)
6. Bielecki, T., Kumar, P.R.: Necessary and sufficient conditions for a zero inventory policy to be optimal in an unreliable manufacturing systems. In: Proccidings of 25 th IEEE Conference on Decision And Control, Athens, pp. 248–250 (1986)
7. Birdwel, J.G.: On reliable control systems design. Ph. D. Disertation, Electrical Systems Lab., Mass. Inst. Technology, report no. ESL-TH-821 (1978)
8. Birdwel, J.G., Castanon, D.A., Athans, M.: On reliable control system designs. IEEE Transaction on Systems, Man and Cybernetics 16(5), 703–711 (1986)
9. Boukas, E.K., Liu, Z.K.: Production and maintenance control for manufacturing systems. IEEE Transactions on Automatic Control 46(9), 1455–1460 (2001)
10. Czornik, A., Swierniak, A.: On controllability with respect to the expectation of discrete time jump linear systems. Journal of the Franklin Institute 338, 443–453 (2001)

11. Czornik, A., Swierniak, A.: Controllability of discrete time jump linear systems, Dynamics of Continuous. Discrete and Impulsive Systems Series B: Applications and Algorithms 12(2), 165–189
12. Czornik, A., Swierniak, A.: On controllability of discrete time jump linear systems. In: Procedings of 4th World CSCC, Athens, Greece, pp. 3041–3044 (2000)
13. Czornik, A., Swierniak, A.: Controllability of continuous time jump linear systems. In: Proc. of 10th Mediterranean Conference on Control and Automation, CD-ROM, Portugal, Lisbon (2002)
14. Ehrhardt, M., Kliemann, W.: Controllability of linear stochastic system. Systems and Control Letters 2, 145–153 (1982)
15. Florentin, J.J.: Optimal control of continuous-time Markov stochastic systems. Journal of Electronics Control 10, 473–481 (1961)
16. Ji, Y., Chizeck, H.: Controllability, observability and discrete-time markovian jump linear quadratic control. International Journal of Control 48(2), 481–498 (1988)
17. Ji, Y., Chizeck, H.J.: Controllability, stability, and continuous-time Markovian jump linear quadratic control. IEEE Transactions on Automatic Control 35, 777–788 (1990)
18. Kazangey, T., Sworder, D.D.: Effective federal policies for regulating residential housing. In: Proc. Summer Comp. Simulation Conf., Los Angeles, pp. 1120–1128 (1971)
19. Kalman, R.E.: On the general theory of control systems. In: Proc. First IFAC Congress Automatic Control, Moscow, Butterworths, London, vol. 1, pp. 481–492 (1960)
20. Klamka, J., Socha, L.: Some remarks about stochastic controllability. IEEE Transactions on Automatic Control 22, 880–881 (1977)
21. Klamka, J., Dong, L.S.: Stochastic controllability of discrete time systems with jump Markov disturbances. Archiwum Automatyki i Telemechaniki 35, 67–74 (1990)
22. Krasovski, N.N., Lidski, E.A.: Analytical design of controllers in systems with random attribute. Parts I-III, Automation and Remote Control (Part I) 22, 1021–1027, (Part II) pp.1141–1146, (Part III) pp.1289–1297 (1961)
23. Mariton, M.: Jump Linear Systems in Automatic Control. Marcel Dekker, New York and Besel (1990)
24. Montgomery, R.C.: Reliability considerations in placement of control systems components. In: Proc. AIAA Guidance and Control Conference, Gatlinburg (1983)
25. Rosenbrock, H.H., McMorran, P.D.: Good, bad, or optimal? IEEE Transactions on Automatic Control 16(6), 552–554 (1971)
26. Siljak, D.D.: Reliable control using multiple control systems. International Journal on Control 31(2), 303–329 (1980)
27. Siljak, D.D.: Dynamic reliability using multiple control systems. In: Proc. 2nd Lawrece Syphosium on Systems Decision Sciences, Berkley (1978)
28. Swierniak, A., Simek, K., Czornik, A.: Fault tolerant control design for linear systems. In: 5th National Conference on Diagnostics of Industrial Processes, pp. 45–50. Tech. Univ. Press of Zielona Gora, Poland (2001)
29. Swierniak, A., Simek, K., Boukas, E.K.: Intelligent robust control of fault tolerant linear systems. In: Proceedings IFAC Symposium on Artificial Intelligence in Real-Time Control, pp. 245–248. Pergamon Press, Malaysia (1997)
30. Sworder, D.D., Rogers, R.O.: An LQ- solution to a control problem associated with solar thermal central receiver. IEEE Transactions on Automatic Control 28, 971–978 (1983)
31. Willsky, A.S., Levy, B.C.: Stochastic stability research for complex power systems. Lab. Inf. Decision Systems, Mass. Inst. Technology, report no. ET-76-C-01-2295 (1979)
32. Zabczyk, J.: Controllability of stochastic linear systems. Systems and Control Letters 1, 25–31 (1981)

Part II
Information Management Systems and Project Management

Chapter 7
General Description of the Theory of Enterprise Process Control

Miroslaw Zaborowski

Abstract. The theory of Enterprise Process Control is a system of notions and relationships between terms referring to enterprise management and process control systems. The universal model of organizational and functional structures of such systems has been presented in the paper. It has been shown that these structures are tightly related to the structure of data flow between decision procedures in information systems that support business process control in enterprises.

7.1 The Purpose and the Scope of the EPC II Theory

The purpose of the EPC II theory (theory of enterprise process control) is organizing terms concerning integrated enterprise management and control systems to improve designing and teaching on such systems. Issues of analysis of mutual relationships between its facts (which is a substance of each theory [3]), have been presented in the form of sets of entities and relationships between them [16,17,18,19]. These relationships have been defined as subsets of Cartesian products of the proper sets and as equivalent to them associations in class diagrams of the UML language [13]. The acronym EPC II is used instead of the abbreviation EPC to avoid confusing Enterprise Process Control with Event-driven Process Chain, which is a well known method of modeling process structures [4].

The EPC II theory may be applied to describing control of every production process in enterprises, including continuous and discrete manufacturing processes, service processes, as well as data processing for customer needs. The application area of the EPC II theory also encompasses management of auxiliary processes, such as overhauls, repairs, trainings and the like, as well as all internal administrative

Miroslaw Zaborowski
Academy of Business in Dabrowa Gornicza, Department of Computer Science
ul. Cieplaka 1c, 41-300 Dabrowa Gornicza, Poland
e-mail: mzaborowski@wsb.edu.pl
http://www.wsb.edu.pl

A. Kapczynski et al. (Eds.): Internet - Technical Develop. & Appli. 2, AISC 118, pp. 85–93.
springerlink.com

processes. The first part of the EPC II theory is description of the framework EPC II system, which is a universal reference model of integrated business process control systems. The tentative thesis of the EPC II theory is the statement, that every management or process control system, irrespective of size and trade of an enterprise in which it is implemented, may be replaced, retaining all its functions and data, with a corresponding EPC II system, whose structure is the same as for the framework EPC II system [19]. The second purpose of the EPC II theory is analysing its compatibility with real business process management systems and with corresponding standards of information processing systems for management and control in enterprises (MRPII/ERP [8,10], ISA-95 [1,5,12], IEC 61499 [9,14], BPMN [2], YAWL [6], WFMC standard [15]).

7.2 Multilayer Structure of the Framework EPC II System

From the viewpoint of management engineering and automation the framework EPC II system is such a net of control units and information places in which control units do not communicate directly, but through information places, and information places are separated by control units. Control units (SJ in the figure 6.1) are arranged in functional layers, whereas information places - in information layers, which separate the functional ones. Among information places one can distinguish activity places (MA in the figure 6.1), which store information on functional subsystems and their activities, and resource places (MR in the figure 6.1), where information on resources, which are used, consumed or produced by functional activities, is written down. In the highest layer there is only one control unit - the supervisory unit of the overall organizational system, which encompasses an enterprise and its environment. In the lowest layer control units are functional units of basic control objects.

Neighboring functional layers are merged into organizational levels, whose control units have similar scope of influence and definite frequencies of information acquisition and decision making. In a typical case an EPC II system has five levels:

- the system comprising the supervisory unit of the overall organizational system
- the overall organizational system, which contains the enterprise and its environment: customers, suppliers, natural environment, banks and the like
- the enterprise plants, which contain their departments
- the organizational cells, which contain their work centers
- the work stations (elementary organizational systems) which contain their direct control subsystems or other elementary subsystems.

An enterprise, departments and work centers are working subsystems, belonging to the same levels as their organizational subsystems belong to. Each organizational level, except of the highest one and the lowest one, has four functional layers:

- coordination of working activities in organizational systems performing business processes,

- allocation of working activities to organizational subsystems of particular working subsystems,
- reengineering (structure changes) for organizational systems of lower levels and their business processes
- scheduling orders for executive subsystems (which are also organizational systems of a lower level)

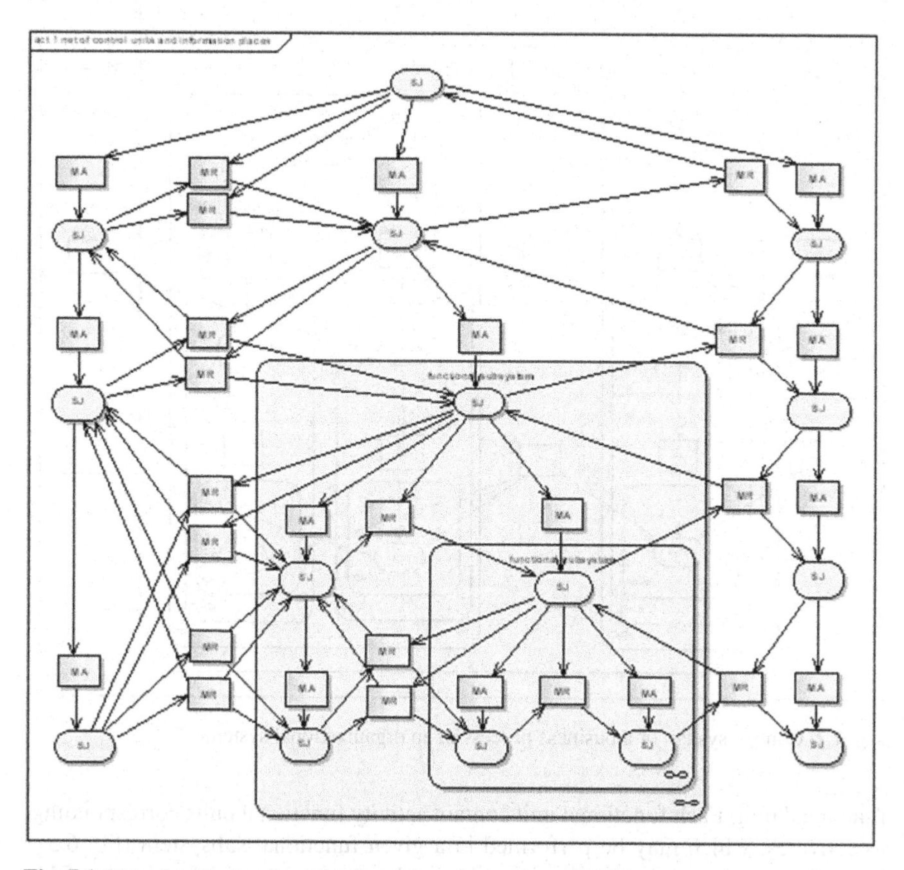

Fig. 7.1 Net of control units and information places in an EPC II system

7.3 Business Processes and Functional Subsystems

Each control unit, except for functional units of basic control objects, sends decisions to control units of lower layers and receives information from them. A system of control units and separating them information places, subordinated to a definite control unit (called the top control unit) or to its subordinate control units, is called a functional subsystem (fig. 6.1). A top control unit, considered as a control object, which receives decisions from control units of higher layers and sends them information, is called a functional unit. Each functional subsystem has exactly one

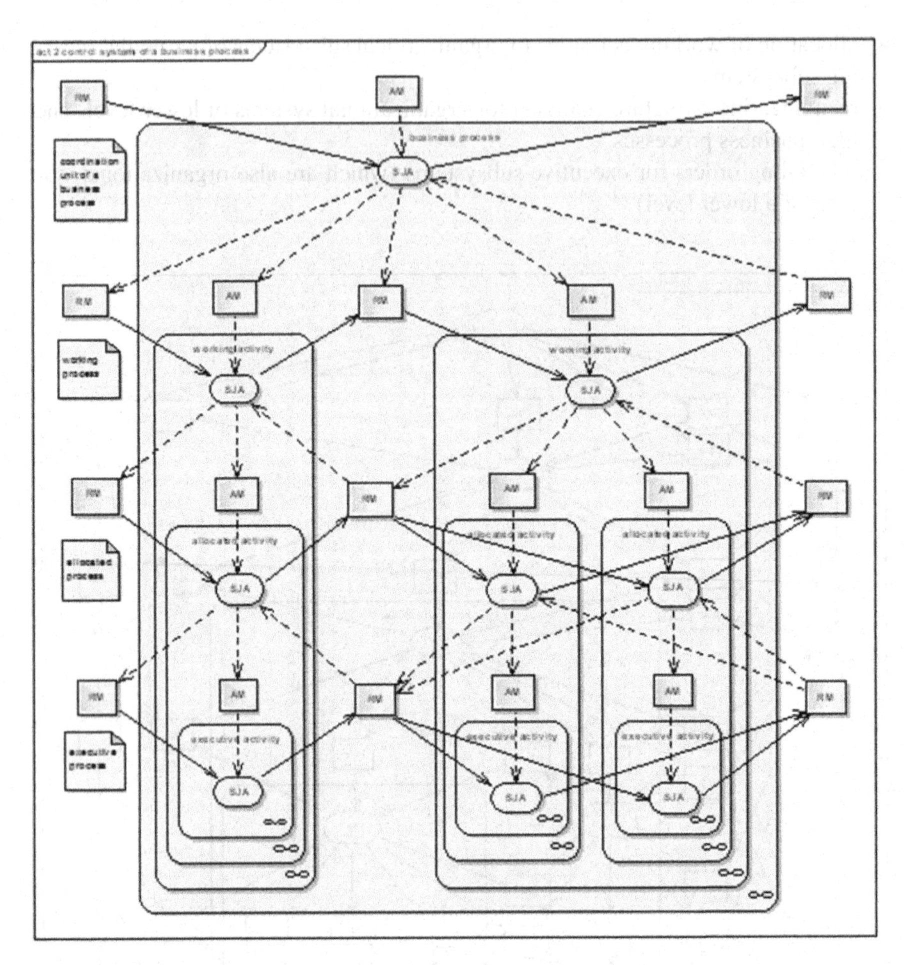

Fig. 7.2 Control system of a business process in an organizational system

functional unit. Each functional unit contain activity functional units corresponding to activities, which may be performed in a given functional subsystem (fig. 6.3). From the viewpoint of control units of higher levels activity functional units (SJA in the 6.2) represent corresponding functional activities. They are called working, allocated or executive activities depending on the layer, to which their functional units belong (fig. 6.2).

Business process is an ordered set of activities and separating them resources, which are used, consumed or produced in these activities. In this definition activities and resources are considered abstractly, as activity kinds and resource kinds [19]. Functional business process is an ordered set of functional activities, represented by their functional units SJA, and separating them located resources (RM in the figure 2.1.2). System business process is an entire process performed in a given

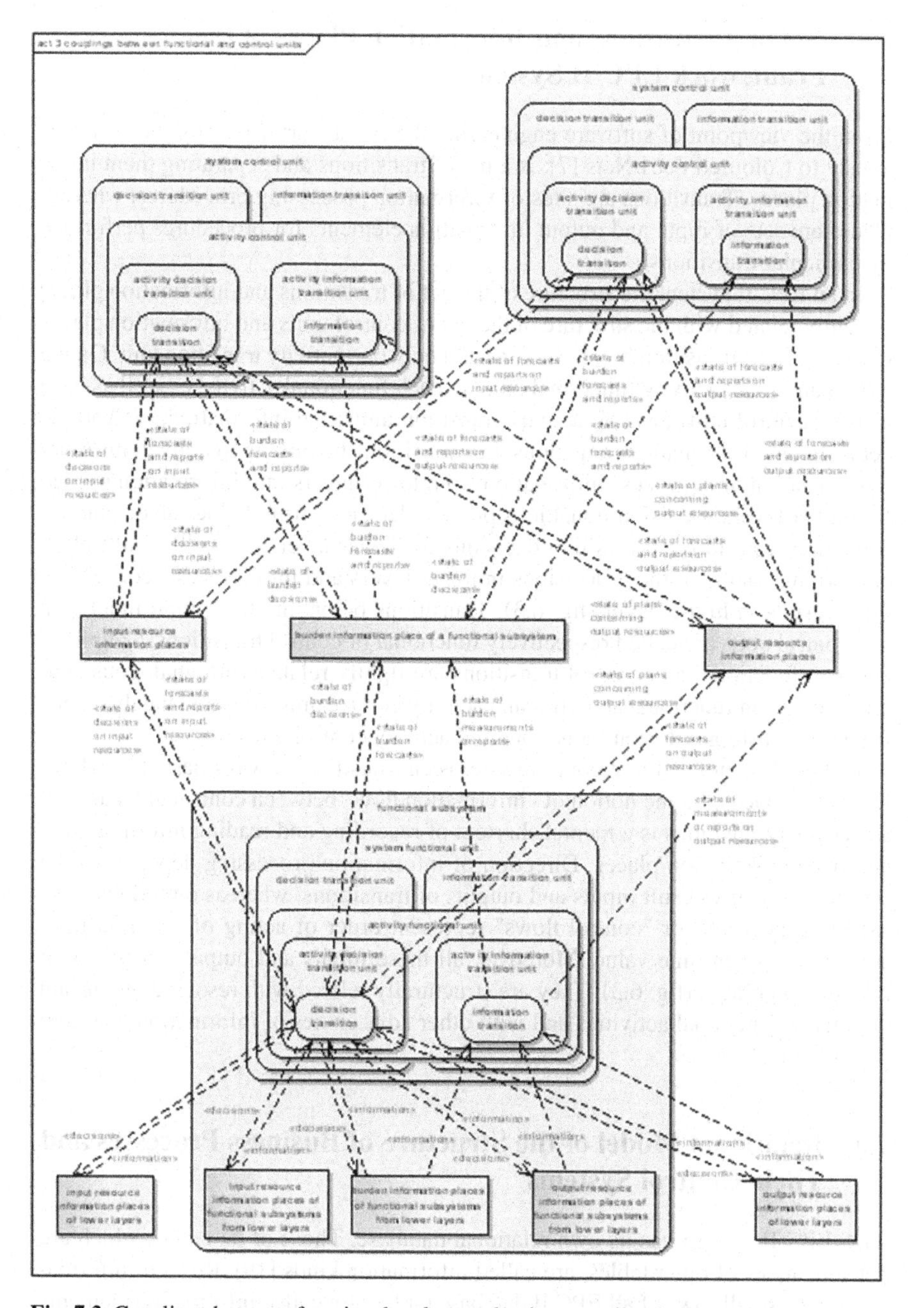

Fig. 7.3 Couplings between functional and control units, as well as between their transitions

organizational system. It is identified by its coordinating control unit (fig. 6.2), which is also an executive activity unit of the higher level.

7.4 Net of Transitions and Information Places of the Framework EPC II System

From the viewpoint of software engineering the framework EPC II system, analogously to Coloured Petri Nets [7], is a net of transitions and separating them information places. Transitions are sites of information processing procedures, whereas places are sets of input and output information elements for procedures performed by particular transitions.

In an EPC II system the structure of the net of transitions and information places is tightly related with the structure of the net of control units and information places (fig. 6.3). Each transition belongs to one and only one activity transition unit. On the other hand, each activity functional unit (so each functional activity, as well as each activity control unit) has two activity transition units: the information one and the decision one. Information transitions, i.e. transitions belonging to a given information transition unit, process information from lower layers into information needed for higher layers. Decision transitions process decisions from higher layers and information from lower layers into decisions for lower layers, as well as into plans concerning results of these decisions in lower layers and into forecasts concerning these results in higher layer (fig. 6.3). Transitions belonging to definite functional and control units are called respectively functional or control transitions. Couplings between functional and control transitions are tightly related with analogous couplings between functional activities and activity control units. It immediately results from relationships between transitions and activity control units.

It should be noticed that couplings between transitions always pass through information places. So the notion of "information flow" between control units as well as between transitions is a mental shortcut of recording and reading information to and from information places. Direction of information processing depends on the structure of cause-result inputs and outputs of transitions, whereas directions of so called "data flows" or "control flows" represent order of acting of different transitions at a given time value. Moreover, all these inputs and outputs are states of information places (fig. 6.3). They are structurally related with resource inputs and outputs of functional activities and with other adjacencies of information places to control units.

7.5 Relational Model of the Structure of Business Processes and Their Control Systems

Each EPC II system has its own relational database. Those of its tables, which are not subclasses of other tables, are called information kinds [19]. Rows of information kinds, so all rows of all EPC II database tables, are called information elements [19]. Information places are separable subsets of the set of all information elements.

The complete list of information kinds, which is the same for all EPC II systems, has been presented, with their key attributes, in [19]. Its three parts, encompass tables designed respectively for storing structural knowledge, information on structure of a given business process control system, as well as information needed for current management and control in a given enterprise. Information kinds of structural knowledge concerning business processes and process control systems may be useful not only in a given enterprise and therefore they can be distributed, perhaps for a payment, among other enterprises. This quality of the framework EPC II system seems to be important in a light of frequently declared striving to a knowledge-based economy.

Attributes of information elements are state variables of a given EPC II system. Each of them may be pointed out by the name of the table, the row identifier and the name of the column or by the number of the corresponding information kind, the number of the information element and the number of the attribute of the information kind. Cause-result inputs and outputs of transitions are determined by their associations with information element attributes.

The framework EPC II system contains the subsystem of business process modeling [20, fig. 6.1], which transforms process diagrams, worked out by users in accordance with rules of the EPC II theory, into relational description of an enterprise structure. The structure of processes, freely modified by users, is tightly related with the structure of couplings between transitions. What is more, it is tightly related with the structure of orders of performing these processes, as well as with the structure of plans, reports and any other information needed for current business process control in a given enterprise. On the other hand procedures, called by transitions from the library [20, fig. 6.1], do not have to change after alterations of process structure. Thus changes of process structure do not involve human interference in an enterprise software. It is a new way of fulfilling the demand of "the third wave" of management methods development that is obliterating the business-IT divide [11]. Obviously, it is always possible, at user's request and independently of process structure changes, to change the content of procedures or change assignment of procedures to transitions. What is more, in the case of structural changes of management methods it is possible to change a selection of active couplings in the set of all couplings of the framework EPC II system.

The framework EPC II system contains the subsystem of business process modeling [20, fig. 6.1], which transforms process diagrams, worked out by users in accordance with rules of the EPC II theory, into relational description of an enterprise structure. The structure of processes, freely modified by users, is tightly related with the structure of couplings between transitions. What is more, it is tightly related with the structure of orders of performing these processes, as well as with the structure of plans, reports and any other information needed for current business process control in a given enterprise. On the other hand procedures, called by transitions from the library [20, fig. 6.1], do not have to change after alterations of process structure. Thus changes of process structure do not involve human interference in an enterprise software. It is a new way of fulfilling the demand of "the third wave" of management methods development that is obliterating the business-IT divide [11]. Obviously, it is always possible, at user's request and independently of process structure changes, to change the content of procedures or change assignment of procedures to transitions. What is more, in the case of structural changes of management methods it is possible to change a selection of active couplings in the set of all couplings of the framework EPC II system.

On the question of integration of ERP, MES and SCADA-PLC systems [1,5,12] the EPC II theory suggests equally radical solutions. Instead of building interfaces between these systems, e.g. such as described by the ISA-95 standard [1], one can build an EPC II system, whose functions encompass tasks of all organizational levels of an enterprise, and in this way remove the need of building these interfaces. Obviously, it removes only structural problems of the interfaces, but not technological problems of information transmission between computers and other technical devices which are included in an integrated system.

7.6 Conclusions

Functional structures of business process management systems may be modelled as nets of functional units and separating them places of information on functional activities and on located resources. On the other hand the functional structure of an information system that supports the management may be presented like a Colored Petri Net with transitions that are sites of data processing procedures and with information places, which are separable subsets of the set of all rows of all tables of the relational database of a given information system. The EPC II theory demonstrates tight relationships between these two structures. Both of them, as well as relationships between them, are recorded in the relational database of a given EPC II system.

References

1. ANSI/ISA-95: Enterprise-Control System Integration. Part 1,2,3,5. (2000-2007)
2. Business Process Model and Notation (BPMN), v.2.0 (2011), http://www.omg.org/spec/BPMN/2.0
3. Encyclopaedia Britannica, http://www.britannica.com/EBchecked/topic/673904/theory
4. Davis, R., Brabander, E.: ARIS Design Platform. Getting Started with BPM. Springer (2007)
5. Gifford, C. (ed.): The Hitchhiker's Guide to Manufacturing Operations Management: ISA-95 Best Practices Book 1.0. ISA Publ. (2007)
6. ter Hofstede, A.H.M., van der Aalst, W.M.P., Adams, M., Russell, N. (eds.): Modern Business Process Automation: YAWL and its Support Environment. Springer (2010)
7. Jensen, K.: Coloured Petri Nets. Springer, Berlin (1997)
8. Landvater, D.V., Gray, C.D.: MRP II Standard System. Oliver Wight Publications (1989)
9. Lewis, R.: Modeling control systems using IEC 61499. Ed. IEE Control Engineering Series 59 (2001)
10. Ptak, C.A., Schragenheim, E.: ERP Tools, Techniques and Applications for Integrating the Supply Chain. The St. Lucie Press (2004)
11. Smith, H., Fingar, P.: Business Process Management. The Third Wave. Meghan-Kiffer Press, Tampa (2003)

12. Scholten, B.: The road to integration. A guide to applying the ISA-95 standard in manufacturing. ISA Publ. (2007)
13. Using UML Part One - Structural Modeling Diagrams. Sparx Systems (2007), http://www.sparxsystems.com
14. Vyatkin, V.: IEC 61499 Function Blocks for Embedded And Distributed Control Systems Design. ISA Publ. (2007)
15. WFMC Terminology and Glossary. Doc. N. WFMC-TC-1011, Issue 3.0 (1999)
16. Zaborowski, M.: The EPC theory. Basic Notions of Enterprise Process Control, Management and Production Engineering Review 1(3), 75–96 (2010), http://www.review.univtech.eu/images/archiwum/nr3/8-zaborowski.pdf
17. Zaborowski, M.: The EPC theory. Couplings Between Transitions in Enterprise Process Control Systems, Management and Production Engineering Review 1(4), 77–92 (2010), http://www.review.univtech.eu/images/archiwum/nr4/9-zaborowski.pdf
18. Zaborowski, M.: The EPC II theory. The Structure of Business Process Control Systems, Management and Production Engineering Review 2(1), 61–77 (2011), http://www.review.univtech.eu/images/archiwum/2011/nr1/9-zaborowski.pdf
19. Zaborowski, M.: The EPC II theory. Data Structure in Enterprise Process Control Systems, Management and Production Engineering Review 2(2), 53–70 (2011), http://www.review.univtech.eu/images/archiwum/2011/nr2/7-zaborowski.pdf
20. Zaborowski, M.: Scope of Applications of the Framework Enterprise Process Control System. AISC, vol. xx. Springer, Berlin (2011)

Chapter 8
On-Line Technological Roadmapping as a Tool to Implement Foresight Results in IT Enterprises

Andrzej M.J. Skulimowski and Przemyslaw Pukocz

Abstract. Technological roadmapping (TR) is a complex business intelligence and strategic planning process. Due to its relatively high costs, duration, mobilisation of human resources and know-how, its hitherto applications were attributed mainly to long-term strategy formation in large corporations and as a tool supporting policy making. This paper will show how the recent development of web technologies and a public availability of foresight results have allowed to provide access to high-level roadmapping even for small and medium-sized enterprises. We will describe an implementation of the roadmapping process in .NET that may serve as an on-line interactive strategic planning and decision support system applying information technology foresight results. The system may be used as a base to solving IT investment planning problems and new product development and market placement problems (NPD-MP) in innovative companies. A web-based analytic machine that supports roadmapping is able to generate optimal technological investment strategies, visualised as multicriteria shortest paths in classical roadmapping diagrams. The system uses the integrated capacity offered by ontologies and semantic networks is a complex collaborative process of knowledge acquisition from heterogeneous information sources, its managing, sharing and processing. An example of the roadmapping process applied to develop a new computer based on IT foresight results will also be presented.

8.1 Introduction

Rapid technological change requires efficient, flexible and dynamic decision making from each market participant. This is a condition for success in business. Strategic

Andrzej M.J. Skulimowski · Przemyslaw Pukocz
AGH University of Science and Technology, Chair of Automatic Control, Decision Science Laboratory, Krakow, Poland
International Centre for Decision Sciences and Forecasting, Progress and Business Foundation, Krakow, Poland
e-mail: {ams,pukocz}@agh.edu.pl

A. Kapczynski et al. (Eds.): Internet - Technical Develop. & Appli. 2, AISC 118, pp. 95–111.
springerlink.com

planning in companies exploits OR-based methods to support the decision-making process such as decision trees, influence diagrams [1], multi-criteria analysis [2], analysis of key technologies and factors, SWOT, and a more recent SWOTC (SWOT with Challenges) analysis [3], etc. These methods are often integrated with ERP systems. However, there is currently a lack of specialized decision-support systems dedicated to specific problems concerning technology transfer and commercialization that are available on the business software market. The direct aim of such systems should be

- to acquire from heterogeneous sources, mainly web-based, collect and process knowledge about the environment (economic, ecological, social, scientific, technical, etc.) in which organizations that implement new technologies operate;
- to develop quantitative and qualitative models of these environmental factors, examine their interdependencies and identifying their dynamics on the basis of previous observations, with the application of optimization and game theory models;
- to classify the causative factors of processes taking place in the environment, which makes it possible to identify the environmental impact and consequences of future decisions to be made;
- to create a general vision of the future (forecasts, trends, scenarios) of the organization and its environment, as well as information concerning the development of specific technologies and products.

The ultimate goal is to develop rules for decision making that optimize criteria relevant to the organization that performs the strategic planning. Depending on how the decision problem is formulated, these rules may take the form of a strategic plan, a list of priorities or a business plan related to the particular technology or product. Despite the development of decision-support methodologies, there is still a lack of technology transfer models suitable for implementation in ERP systems, and very little diversity in the models used in different technical fields. There is also a lack of standardization in the methods used. An extremely important but often overlooked issue is developing algorithms to modify a strategic or other decision-making plan that is dependent on specific future event scenarios for enterprises [4]. Nowadays, every innovative company operates in a highly competitive market, and the exploration of he future socio-economic and technological development is essential for brand building, acquisitions, and strategic market position. Complex and expensive TR and strategic planning processes allowed to fulfill the above presented goals and contributed to the market success of many large enterprises, such as Apple, Microsoft, Motorola etc. This paper addresses an important research area, namely the search for strategic decision support methods that due to new internet technologies, specifically the availability of free information on the web and its filtering and knowledge building would allow to reach that level of quality of strategic planning even for small companies that was previously restricted to major technological multinational companies. We will show that roadmapping is a methodology that allows for integration of a knowledge from different sources, including automatic and semi-automatic exploration of online resources, and expert knowledge. We claim that applying roadmapping-based intelligent decision support systems will strengthen the organizations in building their future. An example of a situation in which an organization should create a strategic plan is searching for new sources of financing for its investment projects and development. Decision makers should thoroughly consider taking certain steps, such as an IPO, acquiring a strategic investor, issuing bonds, a bank loan, mezzanine financing, venture capital funding, etc. For each of these alternatives and their permissible combinations, as well as specific environment scenarios, the impact of the decisions on the current and future situation of the organization should be examined. Various factors must be taken into account, such as market capacity, competition analysis and characteristics of the technology under consideration (its lifetime, the possibility of generating commercial products, the impact of science on the development and resale of technology). Other factors include how technologies depend on legislative conditions, particularly those related to environmental protection, and how decisions affect social phenomena such as outsourcing, reduction or increase in employment, meeting existing or creating new needs. These issues are characterized by a high level of complexity, and most small and middle-sized enterprises (SME) do not have the capacity to analyse them thoroughly and systematically. For this reason, in the development of the computer-aided strategic decision-support systems we have reduced the analysis of external factors to the most important relations from the standpoint of the organization and the particular decision problem in order to concentrate on a thorough study of those factors which most directly influence the decision to be made. Only the information that can be represented in a computer-aided decision support system either quantitatively or as qualitative but verifiable expert judgments and descriptions is processed by the system. Additionally, the decision support methodology for new product development and market placement problem (NPD-MP) presented in this paper makes use of information adopted from technological foresight [4], specifically development trends and scenarios of selected information technologies. The approach here presented has been implemented as a prototype decision support system in the .NET environment.

8.2 The Essence of Roadmapping

One method of supporting complex strategic decision-making processes is called roadmapping ([3],[5], [6], [7], [8], [19]). The essence of roadmapping is:

- breaking down the organization's environment into layers corresponding to interrelated and desirably homogeneous groups of factors, objects and operations
- decomposing the dependencies and relationships within and between layers, where an attempt is made to rank the layers so that the factors of layer n are only associated with the factors of the (n-1)st and (n+1)st layer
- taking into account the time relationships between factors (causal relationships, probabilistic relationships, trends, scenarios, descriptions of dynamics, etc.)
- creating diagrams of dependency factors taking into account the different relationships and their gradation (called roadmaps due to the apparent similarity to

the use of road network maps for selecting the shortest route; also similar to PERT diagrams within each layer the roadmapping concept is not fully reflected)
- identifying key decision points in the diagrams, solving optimization and decision-support problems associated with them.

Another special feature of roadmapping is the simultaneous use of formal and quantitative knowledge, as well as informal expertise and managerial knowledge. The approaches used here are familiar to foresight methodologies, such as Delphi surveys, expert panels, analysing associations (so-called "brainstorming"), etc. Through these methods, a wider and bolder look into the future is possible, as opposed to methods based solely on formal knowledge. This paper presents only a small selection of roadmapping application possibilities, namely Technology Roadmapping, which focuses on decision problems associated with Information Technologies (IT) implementation and commercialization. There are further areas of application for this method in economic, political and social fields ([9], [10]). These will be discussed briefly at the end of Section 5. The concept of roadmapping is currently used in a variety of contexts. In this paper, roadmapping is an interactive, group instrument used for:

- finding the relationships between the individual elements of complex objects related to the technology transfer as well as analysing cause-effect relationships;
- Adapting strategic planning in technology issues;
- Decision support, through well-structured knowledge of the analysed problem.

Further on, we will present the system architecture which makes it possible to implement the above functions in organizations dealing with technology development and application research, technology transfer centers, as well as organizations creating and implementing new technologies. In other organizations, such as financial institutions, government agencies and research institutions, roadmapping can also be used for other tasks, which include forming R§D and innovative strategies. Due to the diversity of scale and content of applications, no uniform methodological approach to roadmapping exists. There are many variants of this methodology, which differ by the number and type of layers, number of analysed factors, type of temporal and causal relationships under consideration, time horizon of decisions, etc. They depend on the problem area, the purpose of the analysis and target group. Therefore, the roadmapping process described below should be treated as a pattern of conduct/blueprint that allows us to create relational structures helpful in modelling and analysing the problem being solved. In this paper, the roadmapping process will therefore be understood as a scheme for implementing interactive decision-support methods for planning and predicting technological development. Due to the diversity of scale and content of applications, no uniform methodological approach to roadmapping exists. There are many variants of this methodology, which differ by the number and type of layers, number of analysed factors, type of temporal and causal relationships under consideration, time horizon of decisions, etc. They depend on the problem area, the purpose of the analysis and target group. Therefore, the roadmapping process described below should be treated as a pattern of conduct/blueprint that allows us to create relational structures helpful in modelling and analysing the problem being solved. In this paper, the roadmapping process will therefore be understood as a scheme for implementing interactive decision-support methods for planning and predicting technological development. Applying technological roadmapping in enterprises can take the following form:

1. modeling the evolution of technologies used by organizations, especially product and process technology;
2. forecasting demand for products and technologies;
3. planning and optimizing strategies that ensure the technological development of organizations.

In Problem (c), roadmapping is used to solve a multicriteria optimization problem. The choice of development strategy is implemented assuming the simultaneous optimization of several criteria such as profit as a function of time (the problem of trajectory optimization), the risk associated with the implementation of specific strategies, as well as the company's strategic position (including its market position). Different organizations can also apply specific additional criteria.

8.3 Formulating a Technological Strategic Planning Problem

A basic strategic problem that can be solved with computer-aided roadmapping techniques is New Product Development and Market Placement (NPD-MP, in short NPD). The company faces the challenge of developing a product that will be competitive in the market. Assuming the technological investment has been made, the following pre-criteria will determine the success of the product on the market:

- time t_0 to product launch (measured as a relative criterion, with respect to the start of implementation activities or to a known or estimated time of launch of similar products by competitors);
- average unit cost of the product *c(k)* in the k-th forecasting period, $[t_k, t_k + \Delta t]$, not including the cost of depreciation of the technology;
- predicted market life, $T - t_0$, where T is the expected date of production completion;
- estimated demand *s(k)* for the product by customers in the *k-th* period, $s(k) := \sum_i s_i\{\rho_i(k), \sigma_i(k)\}$

where *i(k)* is the price of the product on the *i-th* market, and $\sigma_i(k)$ is the estimated product market position index in the i-th market in the period $[t_k, t_k + \Delta t]$, which is dependent on factors such as the degree to which the product meets customer needs and the presence of competing products. A sum is made of all the markets where the product will be sold. Estimating the values of these criteria requires the implementation of product market research, competition analysis and a study on the technologies currently available and expected in the planning period. The latter can be usually accomplished by acquiring results of a foresight exercise [3]. The concept of product used above is a simplification, as a 'product' can also be technology, and in certain cases it can be identified with the technology employed in its production. The product can also be understood here as a set of complementary products manufactured using the same technology, or as a result of the same investment process. The final decision to make the technological investment is dependent on the assessment of the economic parameters of the product throughout its life cycle. Discounted cash flow (Net Present Value) related to the implementation and operation of the new technology is usually taken into consideration as the final criterion:

$$MPV(I,t,d) := C(0) + \prod_{k=1}^{t} \frac{C(k)}{\prod_{1\leq j\leq k}(1+d_j)} \tag{8.1}$$

where: I - the technological investment characterized by cash flow $(C(0),,C(t))$ in subsequent accounting periods, $C(0)$ is the initial investment; t- the number of time units since the beginning of the technological investment until the planned completion of production T, $d = (d_1,...,d_t)$ - the average expected discount rates in subsequent accounting periods (it should be added that the generally-used models with a constant discount rate in situations of highly volatile rates can result in serious financial errors). Cash flows $C(k)$ over period k consist of revenues from sales generated by the investment $C_1(k) := N_1(k) * p(k)$, remaining investment revenue, including revenue from the reinvestment of surplus cash $C_2(k)$, costs of investment $C_3(k)$, fixed production maintenance costs $C_4(k)$, as well as variable costs of production $C_5(k) := N_2(k) * c(k)$ which depend on its size, i.e.

$$C(k) = N_1(k) * p(k) + C_2(k) - C_3(k) - C_4(k) - N_2(k) * c(k), k = 1..t \tag{8.2}$$

All these functions should be treated as random variables with distributions estimated from a sample as well as on the basis of market research and various heuristics. In practice, formulas (1) and (2) apply to expected values, and stochastic analysis reduces to variance analysis or other risk measures. Note that in the criterion (1) we have already taken into account the values of pre-criteria k_0, $T - k_0$,$c(t)$ and $s(t)$. The latter is included in the sales forecast $N_1(t)$. However, criterion (1) can be further extended to include terms related to real options values [12] that can often be identified in IT management problems. That allows for a more adequate modeling of the strategic situation of the organization implementing the new product or technology. When using the real options the demand $s(t)$ can be considered a vector criterion scaled through the valuation of real options. As mentioned above, technological measures of investment risk are further criteria used in the strategic planning process. The following can be used, alternatively or simultaneously:

- a variance or semivariance of $NPV(i,t,d)$ (see (1) - (2));
- the probability of a loss of liquidity in the organization during the investment, determined by analysing cash flow $C(k)$;
- the probability of achieving the technological investment goals, which affects the deviation of NPV from the value determined by market analysis.

In addition, in supporting decisions related to technological planning, commercialization of technology and development of production, objectives and strategic criteria are taken into account. These include conformity of the investment with the strategic objectives of the company, conquering new markets, weakening competitors and achieving a competitive advantage. Other criteria can include degree of achievement of another strategic objective, which may be to gain strategic customers, etc. These indicators can be in the form of reference sets [2]. The need to take into account multiple criteria simultaneously transforms the problem (1)-(2) into a multi-objective optimization problem. However, the assumption that in the problem of production planning criteria (1) or (2) are optimized as well as criteria related to risk, treated as a function of final time T, leading to the formulation of a discrete dynamic multi-objective optimization problem:

$$[J \ni I \rightarrow (NVP(I,t_1,\cdot),...,NVP(I,t_2,\cdot))] \rightarrow max \tag{8.3}$$

where J is the set of allowable technology strategies (under consideration), t_1 and t_2 correspond to the minimum and maximum permissible deadlines for the settlement of the investment to be made. In problem (3), the discount rate is not a decision-making variable, but an external random variable whose values are estimated in the forecasting process. Note also that problem (3) is equivalent to the problem:

$$[J \ni I \rightarrow (NVP(I,t_1,\cdot),\frac{C(t_1+1)}{\prod_{1\leq j\leq t_1+1}(1+d_j)},...,\frac{C(t_2)}{\prod_{1\leq j\leq t_2}(1+d_j)})] \rightarrow max \tag{8.4}$$

where the precriteria are related to criteria in a more intuitive way. Finally, a multi-objective optimization problem associated with a choice of technological strategy can be written as

$$\begin{aligned} [J \ni I \rightarrow (NVP(I,t_1,\cdot),...,NVP(I,t_2,\cdot))] &\rightarrow max \\ [J \ni I \rightarrow (R(I,t_1,\cdot),...,R(I,t_2,\cdot))] &\rightarrow max \\ [J \ni I \rightarrow (S(I,t_1,\cdot),...,S(I,t_2,\cdot))] &\rightarrow maxx \end{aligned} \tag{8.5}$$

where R is the measure of risk and S a valuation of the strategic position of a company concerned. Determining the precriteria values, the relationships between precriteria, as well as cash-flow values, and consequently, R and S as formal criteria in problem (5) is generally not an easy task. It requires an examination of the relationship between technologies, products, sales markets, as well as market, economic and political environment forecasts, and the development and application of technological forecasting. All the elements and factors are interrelated, while in practical problems the number of relationships is very high, and their nature is usually heterogeneous: deterministic, stochastic and fuzzy. The data gathering and fusion processes are complex as well as qualitative preference information must be obtained from the decision maker in problem (5), from other decision makers in the same enterprise as well as from external experts.

Problem (5) is the main theoretical basis for roadmapping applications in NPD problems. Section 4 will present an example related to the manufacture of a portable computer relating it to the technological trends in the development of electronic components. Another example of a similar roadmapping process applied for the NPD-MP problem was described in [3].

8.4 Applying Roadmapping in Technological Investment Planning

The analysis described above can be described algorithmically. Both formal and non-formal methods of acquiring knowledge about the environment of problem (5) are applied. This takes into account estimates of C(k), R, S, and the rules of technology strategy choice compromise in (5). In practice, roadmapping requires the quantitative and qualitative analysis of a large number of events in various fields, which need not be related directly to the development of the product itself. The analysis may cover an entire sphere of business activity. This methodology may be used to solve problems such as choosing a technology investment strategy that maximises future benefits (financial, knowledge, human resources), but where resources may be limited. A general roadmapping process scheme for the NPD-MP problem (1)-(2) is shown in fig.8.1. The decision problems and their place in the roadmapping process for product technology are shown in Fig.2. The diagrams constructed in the roadmapping process are a projection of alternative visions of present and future linkages in the important areas for the activity of an organization. The primary area - and also one of the layers in the roadmapping diagram - is a set of relationships related to technologies. During the analysis, causal relationships between these areas are defined. The resulting roadmap is an essential element in creating a medium- and long-term organizational strategy. It primarily helps decision makers (technology administrators) to identify and select of the best alternatives. It should be stressed that any action taken during the roadmapping process involves forecasting, and analysing the dynamics is an essential component of roadmapping.

Different roadmapping applications require a different level of detail, when the relationships within certain layers or groups dominate quantitatively the external relationships between layers. What may be desirable is a hierarchical decomposition of the whole process by treating the roadmap as a hypergraph and analysing its individual levels. The culmination of the roadmapping process is an analytical report, which is based on a diagram analysis and contains recommendations for decision makers. A sample diagram created for a manufacturer of notebook computers is shown in fig.8.3. This example is not intended to provide a whole range of products and technologies, which computer manufacturers generally have, but is merely to illustrate the basic ways of building dependencies and identifying objects. To analyse the NPD-MP problem, where the product is notebook computers, and the sponsor is a company is proposing to expand its production, five layers have been selected:

- technology
- products
- market and business environment factors
- development opportunities
- research related to IT and electronics.

The strategic business planning process proposed here runs in three phases: *Phase 1. Preliminary activities* These activities involved preparing data necessary to initiate roadmapping. The scope of the project together with its boundaries and objectives

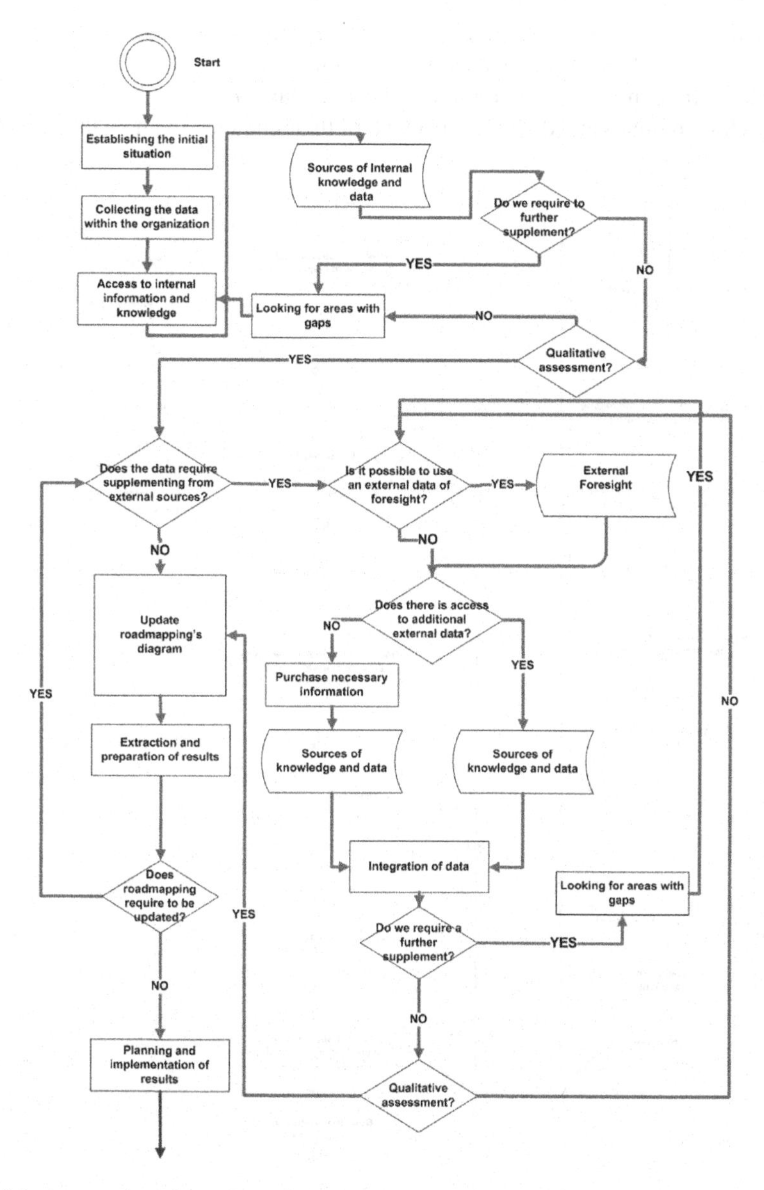

Fig. 8.1 A general roadmapping scheme

was also defined. In the example below, roadmapping is used to select the manufacturer's strategic activities over a five-year horizon, including determining research areas, marketing strategy objectives, acquiring know-how, etc. *Phase 2. Roadmap diagram development.* As part of this phase, a diagram will be constructed by:

1. isolating of five classes of modelled objects: technology, products (which in this example initially leads to the analysis of 10 bipartite graphs)
2. studying the remaining structural links between layers
3. studying timelines and directions (trends) of their changes and how they develop

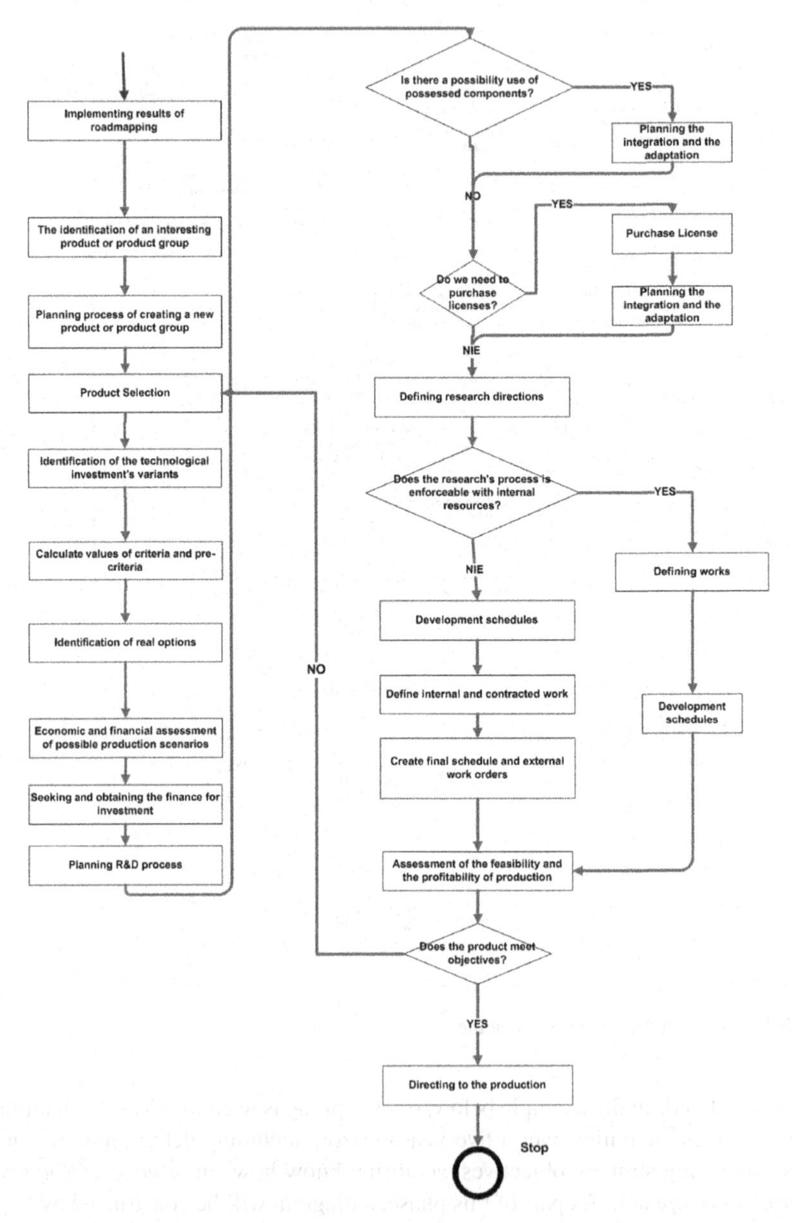

Fig. 8.2 The scheme of the decision-making process in an organization using roadmapping methodology

Phase 3. Result implementation and complementary activities In the IT and electronics industries, which are the main playground for the methods here proposed, technological progress is accelerating at a rapid pace, and the technology life cycle is short. In the example below, the time horizon is 5 years, which means a relatively very short life cycle for the individual products. Periodic updates are therefore necessary within the developed schedule. In this case, measures for upgrading and improving are proposed in the same periods of the year. The aim of these actions is to ensure that roadmapping results are of adequate quality and recommendations are accurate. Finally, the recommendations are presented to decision makers - the diagram is converted into specific actions to be taken in order to achieve the roadmapping objectives. Roadmapping is a cyclical process and therefore it should allow for active monitoring of the reachability of goals in accordance with the whole procedure. This requires the operation of the system in a continuous production cycle as a part of an active environment.

8.5 An Application of Computer Assisted Roadmapping Support System to It Planning

Implementation of the functionality of DSS requires solutions to problems (4)-(5), which leads to the main outputs of the roadmapping process that appear as:

- diagram roadmaps,
- report with assessment scenarios and presentation of the best of them in accordance with users' preferences as well as
- key decision points and scenario bifurcations marked on the diagram.

The main advantage of implementing the technological roadmapping process as a web-based decision support system shows in Phase 2 above, after the grouping of objects. Focus groups using a collaborative internet environment distinguish then objects within layers and show the relationships between them and the objects in other layers. This is an interactive process, and the intermediate results are assessed, discussed and improved. In this phase, a report and roadmap are compiled together with recommendations for decision makers.

The roadmapping diagram shown in Fig.8.3 includes cause-and-effect relationships for objects of individual layers. During the roadmapping process, two types of linkages are identified: within layers and between layers. For example, relationships between objects in the layer "Products" depend chronologically on the introduction of individual products on the market, taking into account depreciation of the technology and marketing expenditure. In the example shown in Fig.8.3, the following relationships within layers are defined:

- Scientific research - the development of scientific research (a consequence of timelines and research results, the evolution of the characteristics of research fields),
- Technology - the availability of technology as a result of research processes, the purchasing of a license and key components from external suppliers,

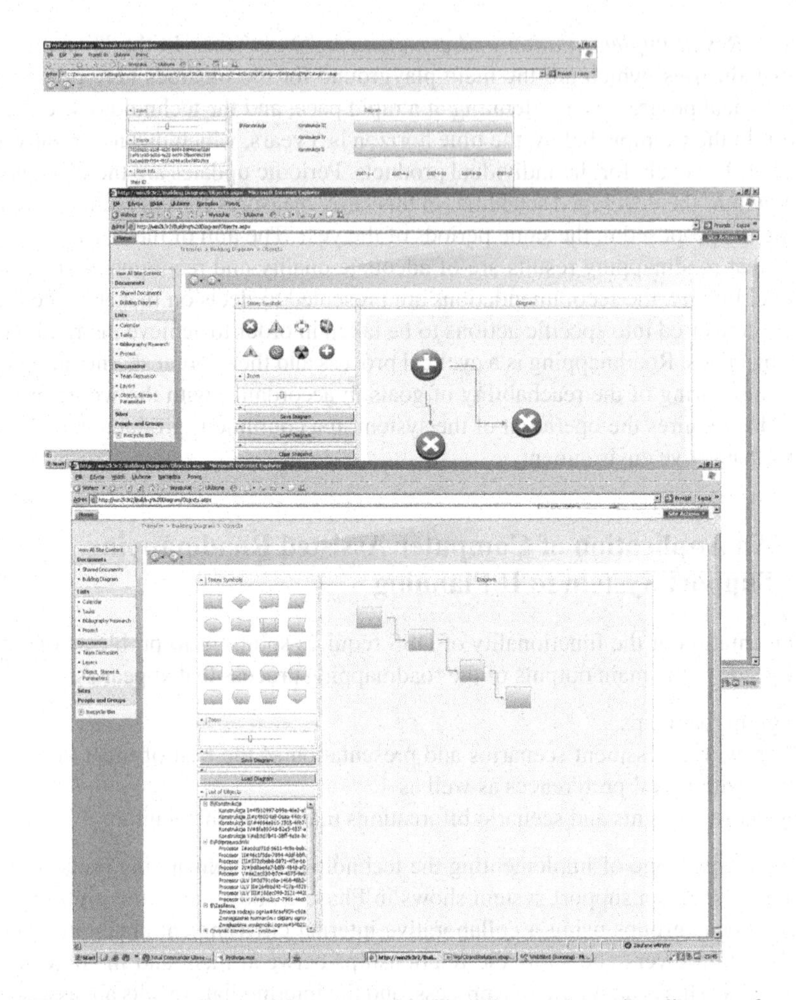

Fig. 8.3 Sample screenshots with an example of building roadmapping diagrams for notebook computers production

- Products - marketed products and their substitutes
- Development opportunities - identification and evolution of developmental factors for the organization,
- Business factors - identification and evolution of business and market factors.

The set of relationships between layers defines the various contexts for the objects of one layer. For example, the product can be analysed in terms of development opportunities as well as real time profit. In a similar way the Scientific Research layer can be viewed from the standpoint of the organization's technological portfolio. In turn, the technology at the disposal of the organization can be used to design and manufacture future products. fig.8.4 shows several relationships to illustrate this process. For example, relationship (1) in fig.8.3 indicates the development of technology

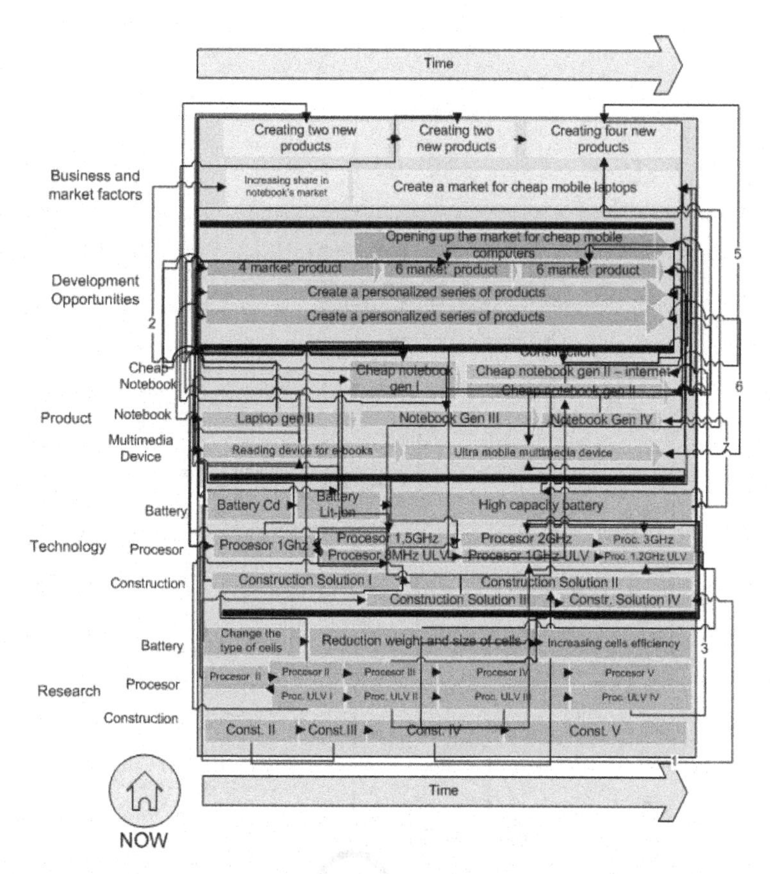

Fig. 8.4 A sample roadmap for a manufacturer of notebook computers

"Construction IV" enabling the creation of a class of casing used in the new product series. The layer "Technology" includes objects from the technological organization portfolio. These objects are turned into real marketable products (see relationship (7)).

In the above roadmapping process, the identification of real options may also occur ([11], [12]). The difficulty in this case lies in the fact that options cannot generally be separated as objects in separate layers, but they can also correspond to relationships between objects in the same layer. These options can be connected using a single technological solution created for a specific product in a few other products in the same category for different market segments. For example, a 2.8 Ghz processor can be used in several different types of devices. An example of options separated in individual layers can be seen in relationships (2), (5) and (6) shown in fig.8.3. The creation of applications implementing the Roadmapping's functions requires developing a new approach to DSS. DSS architectures presented in this paper require new development tools, embedded in a web environment. To build the system we have used the available Microsoft technology: Sharepoint Services as

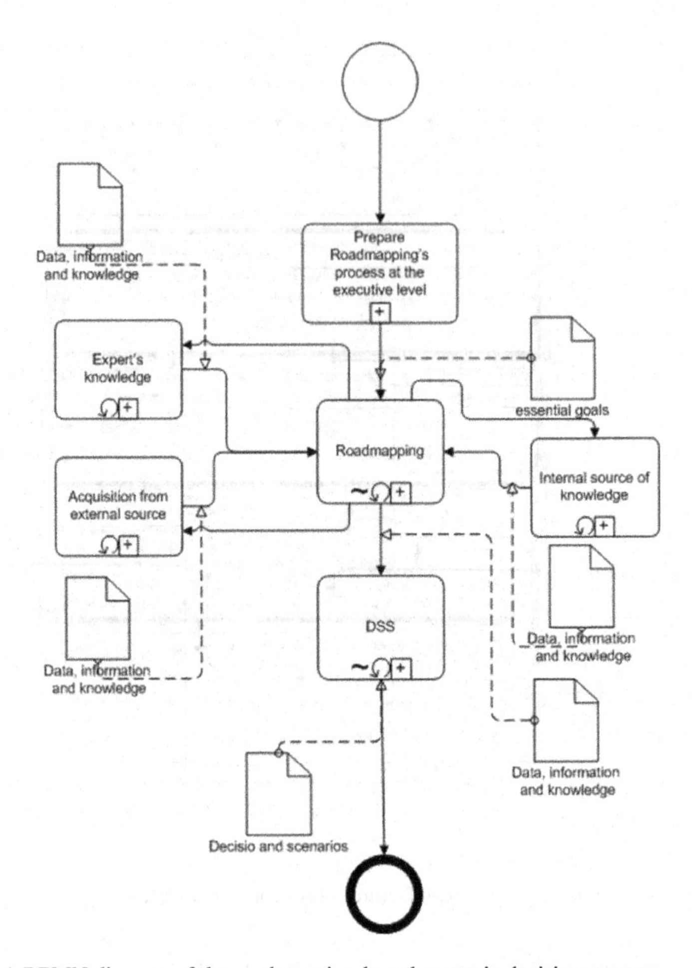

Fig. 8.5 A BPMN diagram of the roadmapping-based strategic decision support

collaboration software for the organization as a programming framework, Microsoft SQL Server, Silverlight, WPF.

The quantity and variety of collected data, information and knowledge through the DSS system is large enough to give rise to the necessity of developing new basis for gathering and processing knowledge-bases. Several methods have been developed that are based on intensive use of ontological knowledge-bases . This leads directly to the use of Ontology and Semantic Web technology as the basis of the information carrier in the system [13]. The BPMN diagram for data, information and knowledge transfer in the above system is presented in fig.8.5. The above-mentioned ontological technologies are available as open source license. An example of the basic package used to create knowledge bases in our system is Jena API, which is developed in cooperation with HP Labs Semantic Web Programme. Use of this package provides access to such technologies as OWL, SPARQL and Pellet.

8.6 Final Remarks

Pursuant to the roadmapping process requirements we have developed a web-based system that allows for active sharing of data, information and knowledge in an on-line strategic planning process. Due to the open availability of technological fore-sight results financed from public funds [4] this process can be undertaken even by small technological companies. When used repetitively in the same enterprise, TR allows to re-use previously gathered knowledge, experience and procedures used in earlier strategic decision-making problems and projects. In most cases the following items repeat while building roadmaps: object-oriented structures, economic environ-ment analysis, market research. It is possible to use individual domain knowledge gained in various projects and at different organizations, in particular the outcomes of technological foresight projects carried out at regional, national or international (EU, OECD) level that are usually in public domain. Utilization of this knowledge to solve NPD-MP problems is essential from the viewpoint of time, money and quality of corporate strategies so derived. The above-presented knowledge management ap-proach is based on the general idea of creating dynamic models of all relevant outer systems influencing the company's performance as well as of internal enterprise subsystems under consideration. DSS technologies based on ontologies provide ap-propriate tools for achieving these objectives [17]. To sum up, roadmapping-based strategic decision support can be useful in increasing the quality and efficiency of the decision-making processes in an enterprise. It allows to indicate several feasible sce-narios to achieve equivalent strategic results that are derived from foresight scenar-ios used at the roadmap building and decision analytics stages. Moreover, roadmap-ping yields optimal strategies from the point of view of such performance criteria such as cost (the cheapest solution), time (the earliest milestone for achieving the desired results), and minimal risk. Prospects for improving the strategic position of the organization can also be pointed out within the process, e.g. by an additional spending on promoting environmentally-friendly activities, information on meeting ISO requirements, etc. Formulating precise criteria levels to be achieved, which may be difficult due to high levels of uncertainty, especially about market situation and financial results (income and earnings), is replaced by the optimization task and qualitative assessment of the impact of implementing selected strategies. Consider-ing additional risk-related criteria, such as statistical moments of random variables describing the state of an enterprise and its environment as well as uncertainty con-cerning the optimization criteria values allows for a formal formulation of a dynamic multi-objective optimization problem and elimination of dominating solutions. This is a base to use supplementary preference information to make a decision on strategy alignment. A direct comparison of roadmapping uses and processes is given in [14], [15], [7], [6]. [8], [16]. The technological roadmapping methodology constitutes an effective organizational framework and a new way of creating a knowledge base on economic, social and technological trends and innovative ideas that are drivers of future enterprise development. The possibility of interactively generating scenar-ios in roadmapping-based DSS [4] means that an adequate model of action under different external conditions affecting an enterprise is achieved more quickly. This approach can easily be adapted to a particular area of application, in virtually all innovative organizations. The use of new tools based on internet technologies and on exploring the internet content can considerably increase the potential of TR as a powerful strategic DSS. The capability to networking a practically unlimited number of decision-makers in an enterprise and an immediate algorithmic verification of the consistency of their judgments allows to significantly improve the quality of the decision-making process. The knowledge gathered from the web, when coupled with internal information on company's resources processed in a separate ERP system is particularly important to solving the NPD-MP problem. In this paper we have demonstrated that it is possible to adopt the roadmapping methodology to create flexible strategic DSS applying foresight results with analytic engines available in a cloud while the sensitive company data will be stored in the intranet. In addition, this approach allows the user to partially automate the process of data acquisition, creating models [18] and business scenarios that stand behind strategic technological planning. This can be regarded as a first step towards the construction of future enterprise decision support-systems based on roadmapping techniques, which will minimize the time spent data gathering and structuring and the costs of interacting with experts and processing the expert knowledge.

Acknowledgements. The research presented in this paper has been supported by the project "Scenarios and Development Trends of Selected Information Society Technologies until 2025" financed by the ERDF within the Innovative Economy Operational Programme 2006-2013, Contract No. WND-POIG.01.01.01-00-021/09.

References

1. Howard, R.A., Matheson, J.E.: Influence Diagrams. Decision Analysis 2(3), 127–143 (2005)
2. Skulimowski, A.M.J.: On Multicriteria Problems with Modification of Attributes. In: Trzaskalik, T. (ed.) MCDM 2006, pp. 117–136. Scientific Publishers of the Karol Adamiecki Economical University, Katowice (2007)
3. Skulimowski, A.M.J.: Methods of technological roadmapping and foresight. Chemik 42, 197–204 (2009)
4. Skulimowski, A.M.J., et al.: Scenarios and Development Trends of Selected Information Society Technologies until 2025. 2010 Annual Report, Progress and Business Foundation, Krakow (2011), http://www.ict.foresight.pl
5. Groenveld, P.: Roadmapping integrates business and technology. Research-Technology Management 40, 48–55 (1997)
6. Phaal, R., Farrukh, C.J.P., Probert, D.R.: Technology roadmapping - a planning framework for evolution and revolution. Technological Forecasting and Social Change 71, 5–26 (2004)
7. Phaal, R., Farrukh, C.J.P., Probert, D.R.: Developing a technology roadmapping system. In: Portland International Center for Management of Engineering and Technology, PICMET, Portland (2005)
8. Phaal, R., Farrukh, C.J.P., Probert, D.R.: Strategic roadmapping: a workshop-based approach for identifying and exploring innovation issues and opportunities. Engineering Management Journal 1(19), 3–12 (2007)

9. Willyard, C.H., McClees, C.W.: Motorola's technology roadmapping process. Research-Technology Management, 13–19 (September-October 1987)
10. Ioannou, C.A., Panagiotopoulos, P., Stergioulas, L.: Roadmapping as a Collaborative Strategic Decision-Making Process: Shaping Social Dialogue Options for the European Banking Sector. World Academy of Science, Engineering and Technology 54, 770–776 (2009)
11. Copeland, T., Antikarov, V.: Real options: A practitionerâĂŹs guide, New York, Texere (2003)
12. Trigeorgis, L.: Real Options Managerial Flexibility and Strategy in Resource Allocation. MIT Press, Cambridge (1996)
13. Silva, C.F.D., et al.: Semantic interoperability of heterogeneous semantic resource. Electronic Notes in Theoretical Computer Science 150, 1–85 (2006)
14. Holmes, C.J., Ferrill, M.B.A., Phaal, R.: Reasons for roadmapping: a study of the Singaporean SME manufacturing sector. In: IEEE International Conference on Engineering Management (2004)
15. Phaal, R., et al.: Starting-up roadmapping fast. Research Technology Management 2(46), 52–58 (2003)
16. Muller, G.: The Role of Roadmapping in the Strategy Process. Embedded Systems Institute, University of Eindhoven (2007)
17. Skulimowski, A.M.J.: Future trends of intelligent decision support systems and models. In: Park, J.J., Yang, L.T., Lee, C. (eds.) FutureTech 2011, Part I. CCIS, vol. 184, pp. 11–20. Springer, Heidelberg (2011)
18. Skulimowski, A.M.J., Schmid, B.F.: Redundancy-free description of partitioned compÂňlex systems. Mathematical and Computer Modeling 16(10), 71–92 (1992)
19. Petrick, I.J., Echols, A.E.: Technology roadmapping in review: A tool for making sustainable new product development decisions. Technological Forecasting and Social Change 71, 81–100 (2004)

Chapter 9
Software Testing in Systems of Large Scale

Wojciech Filipowski and Marcin Caban

Abstract. Testing plays an important role in the process of application a change to an IT system, no matter if the system or the change are simple or sophisticated. The articles covers general topics on: the role of testing, classes of tests and test planning. It also gives an overview on how different test categories can be fitted into the implementation process and what the typical classes of tests are, that play most important role in the process.

9.1 The Role of Testing

Testing plays quite an important role in the process of application a change to an IT system, no matter if the system is complex or not and the change is simple or sophisticated. The success factors, however, are:

- to chose a proper type of test/tests to the unique situation taking into account: all the possible aspects of the change, the system and the specificity of an enterprise - the client;
- to perform the chosen tests as planned with the agreed level of services when it comes to delivery of bug fixes.

There is one more thing so obvious, that is sometimes forgotten or not taken into consideration seriously enough in the whole rush and tension of the project, that needs to be mentioned here to put more emphasize on planning role and preparation for the testing in the implementation process. The thing can be just turned into one simple question: "What is testing for"? And the answer is on one hand - obvious, and on the other hand - difficult to give, since when one tries to explain it, there are more and more arguments coming in, making it hard to drain. Nevertheless, in this article

Wojciech Filipowski · Marcin Caban
Silesian University of Technology
Faculty of Automatic Control, Electronics and Computer Science
ul. Akademicka 16, 44-100 Gliwice
e-mail: Wojciech.Filipowski@polsl.pl

A. Kapczynski et al. (Eds.): Internet - Technical Develop. & Appli. 2, AISC 118, pp. 113–119.
springerlink.com

the authors want to summarize the most important reasons to: understand the great need to perform tests in a project, and the necessity to plan them choosing a proper set of tests each time we run a new change in the system (project of implementation).

Now, let us try to give the answer to our question. To describe it, let's define the area of analysis: since faults or errors while developing and implementing a change in an IT system can emerge in every part of the system, no matter if it is the hardware or software. To simplify the deliberation without any harm to the merit, we won't go too deep in the technology, but rather describe the topic from the project management point o view, enabling the authors derive from their experience in the field of project management. First part of the answer is frankly as simple as the question itself: "Testing is the mean to search for, find and correct faults" [1]. This is correct of course, but it is only a part of the complete answer, so we want now to elaborate a bit on it. As we agreed, testing shows bugs. Now: what are bugs, why we want to find them and what kind of bugs are we able to find? Bugs are any undesirable behavior of the system, for instance: faults in functionality, system performing contrary to the requirements or its purpose, poor efficiency, errors in data operations, faults in configuration, communication or integration errors, and so forth. All this: makes the project plan being more difficult to control, causes the project to be less cost effective and more resource consuming, aggravates system security [2] and at the end puts a risk for business. That is why it is so important to find and fix most bugs possible before the system runs in the new version at the client's area of business. Finally, there are different bugs to find, in nearly every area of the system, so to manage the "search for faults" in an orderly manner, we need to plan to perform several sorts of tests, which are introduced below.

9.2 Types or Classes of Tests

There are numbers of types of tests that are performed on different stages of a project. In this section primary classes of tests will be characterized.

Unit Test. Chronologically, during the implementation phase, when engineers change the source code, the Unit Tests are performed. These tests aim at most simple bugs at the lowest level of implementation work e.g. syntax source code bugs, or logical misunderstanding of requirements. Unit Tests scenarios mainly operate within a single library in terms of source code or within limited area of developed functionality to enable most detailed bug search and fix [3].

System Test (SYS). More complex are the System Tests. These operate from a little wider perspective, i.e. system as a whole - perspective. In addition or as a follow-up to the Unit Tests, System Tests explore the whole functionality delivered within the change and that is why they can find bugs in the links between code libraries, layers of the system or in the area of system ergonomics.

Integration Test (INT). While the system test operates in the area of a single system only, in large integrated environments, to check the reliability of the whole multi-system environment, the Integration Tests are performed. Depending on different factors, and thus on the scope of tests, in a large enterprise, Integration tests are the most important and difficult from the project management point of view, phase of testing in the project. Integration tests check the flow between systems and look for bugs that are often quite big issues when found. Bugs in integration tests need to be checked in different systems and most of cases they are not trivial. This test category contains several subcategories, and those most important among them will be described further in the article. Another worth to mention part is that a bug fix for integration test is a result of strong analysis effort in each of the legacy systems of the integrated environment, even though, the bug fix can be implemented in only one system - for example.

Regression Test. When a bug fix is implemented, the patch is a piece of change itself, so especially when the project is complex, it is a good practice to perform a regression test after applying the bug fix patch. Regression is a way to make sure that when we fix one bug, we don't spoil another part of code. Here one thing is worth to mention: regression is as useful as costly, and resource consuming, so a regression test scenario can't check the whole system again, but covers only the "places" lying in the nearby of the bug-fix patch (in terms of functionality or business process). Another important thing: regression is recommended not only as a part of system test but also in other main types of tests (those are described in the article, e.g. integration test, efficiency test).

Link Test. This group is a classic subcategory to the Integration tests, it is the initial part of Integration Tests in fact. During this part of tests simple connection tests are performed to check and enable the system-to-system communication, and basic process flow is tested rather than full business process scope. This test category is usually performed prior to a massive part of testing e.g. Integration Tests.

Efficiency Test (EFF). Depending on the scope of change implemented in the system, as well as recommendation from the part of the project team responsible for the efficiency, the Efficiency tests check the performance of different parts of the system: from storage infrastructure, layer to layer flow performance, user interface responsiveness, through the overall business process performance. This class of tests can be part of either system or integration tests, or can be planned as a separate phase of tests. It can also be automated using different methods including heuristic ones [4] with dedicated data set prepared (the data preparation activity can be automated too, what is explored more in detail in [5] or manually performed, depending on the need and eventually scenarios composed and tools used.

Client Acceptance Test (CAT). This is the class of tests which client or user team is responsible for. It is usually planned as the last phase of tests executed in the development part of the implementation process, just before the procedure of transfer the solution (code) to the production system environment. The test scenarios cover a wide range of system functionality, business processes, ergonomics and partly efficiency (especially the user interface responsiveness). Often with this phase of tests, contract acceptance events are linked and dependent from. In other words, from the contract viewpoint, this test category is the main decision making milestone towards the acceptance of the whole solution.

Version Upgrade Test. After positive closing of acceptance tests, the final part of the implementation process begins. During this phase some sort of tests can also be performed. Among them are version upgrade tests, which project team together with maintenance crew are responsible for. These tests are performed on the pre-production system environment and help to check the proper process of the version upgrade procedure.

Trial-run Test. This is the test category that is performed as the last test during the implementation process. It occurs short after upgrading the production environment and just before the client (or business) starts using the new functionality. It is quite an interesting part, since it happens uniquely on the real - production environment, real processes are run, but only data is not quite real, because data is a specially prepare set for that occasion. Anyway, short after tests are performed - depending on the system specificity - the data is either cleaned or other method is used to disable the influence of testing activity on the production environment, so that real business data and processes are not spoiled.

9.3 Test Planning and Choosing the Proper Test Set

Testing occurs as part of the implementation process, so each part of tests needs to be planned and placed in a proper section of the process. The importance of understanding the relations between the project phases and test types was already appreciated in scientific descriptions [6]. Let's take a brief look at the flow of an exemplary implementation process: it contains of standard stages according to the methodology a project is run with compliance to.

Figure 1. shows the flow of it with indication of chosen test set mapped on the stages of the process.

As we can read from the illustration above the usual usage of certain test categories is linked to certain implementation stages. Thus, Unit Tests, System Tests and partially the EFFs are performed when the code is changed, but what is worth to mention here is that the Efficiency tests realized on that stage is often automated rather than executed manually.

As for the Testing stage, this part is so to say "reserved" for testing, what means, that during this stage of the implementation process most effort is put on testing. Among the test categories here are mainly: SYS, INT and EFF but even if the System and efficiency test were performed earlier, they can be also executed here, but with enhanced scenarios or as a follow-up.

After the Test stage of the implementation process, next one occurs. It is called Implementation from the fact, that code is transferred from test environments to the production one. Short before the transfer the Client acceptance tests are executed. They are of both: technical and formal importance.

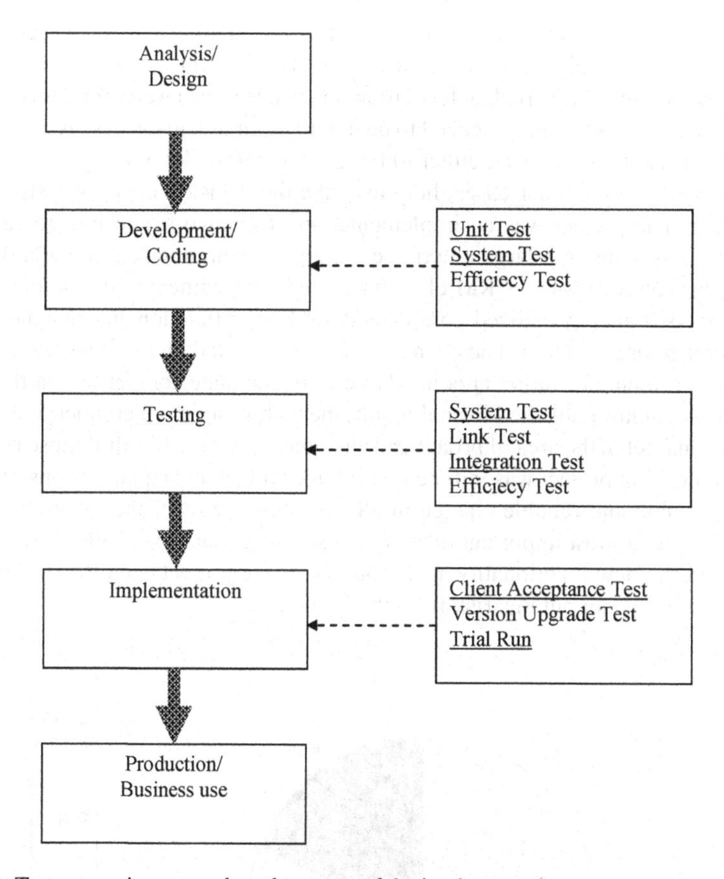

Fig. 9.1 Test categories mapped on the stages of the implementation process

In technical sense, the tests proof the quality and reliability of the solution, as well as the user interface ergonomics, if applicable.

Formal meaning of the CAT is that usually, from the contract point of view, the approval of the acceptance protocol, which is dependent on the positive closing of the tests.

This shows how tests planning is important during the course of the project. Each test category needs to be performed in a proper time, so the application code quality needs to be improved gradually and in a correct order providing more improved code for more complex tests.

Equally important as test planning is the selection of the proper test set. To help understanding which tests are most important we can use the lessons learned technique to review how previously conducted and successfully finished projects were planed in the context of testing. Figure 2. shows the share of main test categories on the base of the overall test scope. This diagram is derived from the analysis of the real project plan where the mentioned test categories had been chosen and executed during the project accomplishment.

Now we need to describe a little bit the project surrounding and system specificity what undoubtedly influenced the choice of test types in this project. Generally, to understand the role of individual test group one needs to be aware of all the internal and external factors that project need to deal with as far as a given case is concerned. These considerations can lead either to build a logical model of the project-system relations [7] or, in simpler cases, help to make the decision on what tests should be chosen. The project was an implementation of change to the integrated environment of systems in a large enterprise - a big telecommunication operator. The system environment was a CRM class integrated environment with complex business processes and specialized components to deliver functionality matched with the enterprise organization. The change itself was initiated by the Regulator, which was a government institution appointed to ensure adequate competition in the market. The Regulator published formal requirements forcing more competition on the domestic market in the area of broadband customer services. To fulfill those requirements quite a big project was initiated with huge budget and equal responsibility to deliver a stable and reliable change in all the components of the integrated environment. One of most important factors in ensuring the success in that undertaking was to test the new functionality as thoroughly as the project can effort so that the organization was not put in a risk from the Regulator.

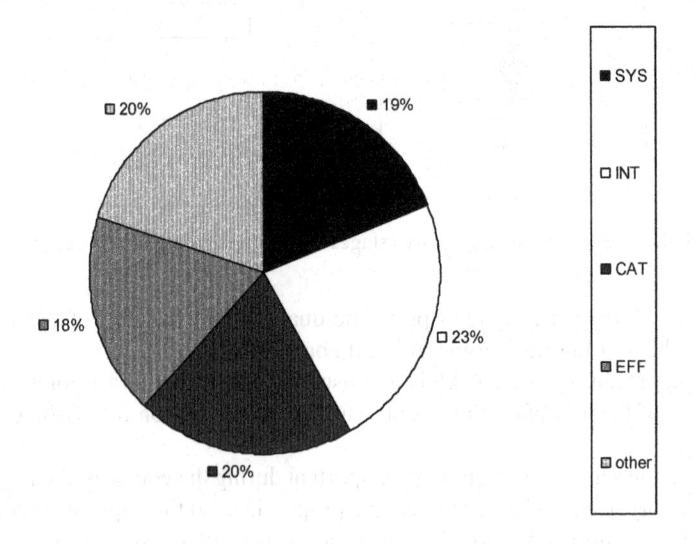

Fig. 9.2 The share of main categories of tests throughout the overall test scope

The result of the choice is shown on the diagram 2, where most important part is how those test categories share the scope of testing to provide the organization with a reliable solution. The chose test categories are: System tests (SYS), Integration tests (INT), Client Acceptance tests (CAT) and Efficiency tests (EFF), while tests grouped in the "other" part of the chart are: Unit tests, Link tests, Upgrade version tests. The numbers there represent effort put on a certain test category in comparison with others. The conclusion from this diagram is: when a proper set of different sorts of tests is selected, each of them is neraly equally important when it comes to improve the tested code.

9.4 Conclusions

There were two general objectives of the article. First one was to name and define the most important test types used in software testing in implementation projects in large companies which operate complex integrated environments. The other was to take a closer look at the problem of estimating the need for a certain test class, than to show how important is to select complementary test set so each part of the set takes its own part of responsibility when it comes to system testing in the name of the compliance with the requirements. The topic, however, seems not to be fully explored, since the perspective chosen, which was quite a general, but otherwise the author wouldn't have been able to fit such a big topic in such a limited volume. All the readers who would like to discuss more on that, are warmly welcome to contact the authors who willingly both: share their experience and build it up by running and taking part in projects with a lot of enthusiasm and commitment.

References

1. Karhua, K., Taipalea, O., Smolander, K.: Investigating the relationship between schedules and knowledge transfer in software testing. Information and Software Technology 51(3) (March 2009)
2. de Vriesa, S.: Software Testing for security. Network Security 2007(3) (March 2007)
3. Link, J., Fröhlich, P.: Unit Testing in Java. How Tests Drive the Code. Morgan Kaufmann (2003) (Imprint) ISBN 13: 978-1-55860-868-9
4. Díaz, E., Tuya, J., Blanco, R., Dolado, J.J.: A tabu search algorithm for structural software testing. Computers and Operations Research 35(10) (October 2008)
5. Sofokleous, A.A., Andreou, A.S.: Automatic, evolutionary test data generation for dynamic software testing. Journal of Systems and Software 81(11) (November 2008)
6. Karhu, K., Taipale, O., Smolander, K.: Investigating the relationship between schedules and knowledge transfer in software testing. Information and Software Technology 51(3) (March 2009)
7. Utting, M., Legeard, B.: Practical Model-Based Testing. Elsevier Inc. (2007) ISBN: 978-0-12-372501-1

Chapter 10
Scope of Applications of the Framework Enterprise Process Control System

Miroslaw Zaborowski

Abstract. The framework EPC II system (Enterprise Process Control) is a universal reference model of integrated management and process control systems for all enterprises, irrespective of their trade and size. Its abilities to work as a skeleton of a company's own software, as an add-on of a purchased management and control system for reducing problems with adapting it to changes of business process structures and as a simulator of business processes and their control systems have been presented in the paper.

10.1 The EPC II Theory

The framework EPC II (Enterprise Process Control) system [7,8,9,10,11] is a universal reference model of integrated management and process control systems. However this model is substantially different than reference models delivered by vendors of ERP (Enterprise Resource Planning [5]) software. Firstly, this is not a trade model, but a universal model. Secondly, it encompasses not only the ERP level, i.e. the management level in an enterprise, but also the level of management in its organizational cells and the level of control inside its workstations. Thirdly, its description was not arisen inductively, as description of functions of offered software in its former applications, but deductively, as the system of notions and relationships between terms concerning information flow in enterprise management and process control systems. This description was named "the EPC II theory" [7,11].

Miroslaw Zaborowski
Academy of Business in Dabrowa Gornicza, Department of Computer Science
ul. Cieplaka 1c, 41-300 Dabrowa Gornicza, Poland
e-mail: mzaborowski@wsb.edu.pl
http://www.wsb.edu.pl

A. Kapczynski et al. (Eds.): Internet - Technical Develop. & Appli. 2, AISC 118, pp. 121–130.
springerlink.com

10.2 The Framework EPC II System as a Skeleton of the User's EPC II Systems

The most natural application of the framework EPC II system is using it for creating a new user's EPC II system (fig. 9.1). For this purpose one should:

1. work out diagrams of business processes with the use of the graphic tool of the framework EPC II system (fig. 9.1),
2. generate, with the use of this tool, the reference functional structure of the business process control system [9] for a given enterprise,
3. remove from this structure those control units and couplings between them, which should not be active according to methods of management and process control, that have been accepted for a given enterprise,
4. introduce such transitions to active transition units [8,9], which are in accordance with accepted methods of management and process control,
5. work out the library of transition procedures, from which procedures may be called by transitions of an EPC II system (fig. 9.1),
6. suplement tables of information kinds and tables of their subclasses with non-key columns [10], designed for data and results of transition procedures,
7. fill in the database of an EPC II system with user's data.

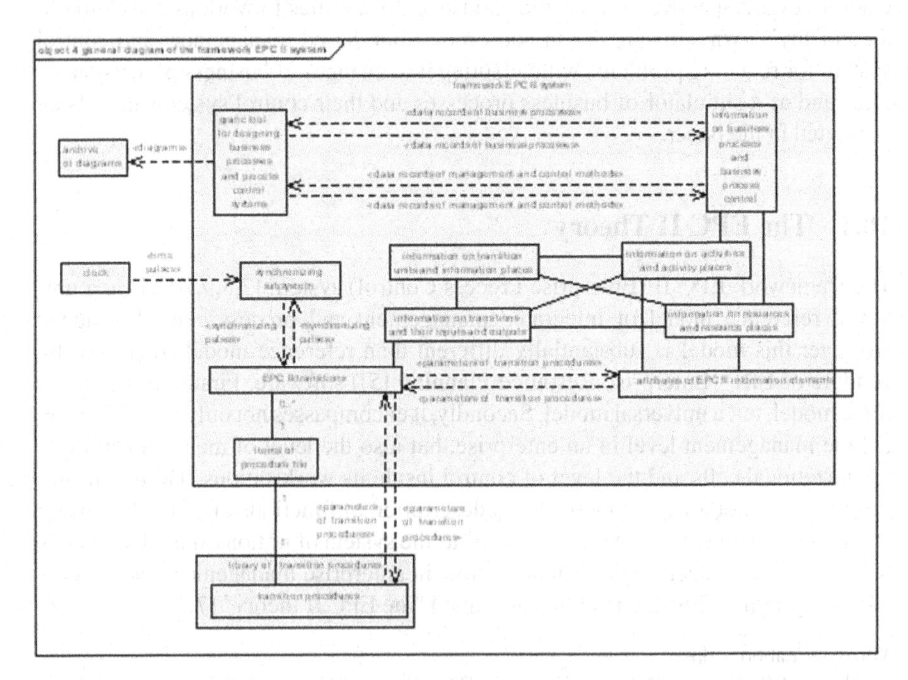

Fig. 10.1 General diagram of the framework EPC II system. Source [8].

The above mentioned phases of building a new EPC II system concern designing and programming the system. For designing it with the use of the framework EPC II system it is enough to precisely know requirements concerning the designed system. The example of the business process diagram is shown in the fig. 9.2, where SP(13, 3, 51) is a system business process [11] of the kind p = 51, performed in an organizational system u = 13 and converting the resource of the kind r =31, monitored at the information place m = 130133 (belonging to the layer j = 3 of the level h = 3), into the resource of the kind r = 51, monitored at the information place m = 11533 (belonging to the same layer) [9]. In the organizational level h = 3 the system business process (13,3,51) is an executive activity of the kind a = 13, performed in an organizational subsystem s = 13 and monitored in the executive functional layer j = 3 [9, 11]. In the executive layer (j =3) of the level h = 2 it is an executive functional business process [11], consisting of two executive activities of the kinds a = 219, a = 211 that are performed respectively in the organizational subsystems s = 41, s = 111 [9]. In this level (h = 2) the border resources r = 31, r = 51, are monitored at the information places m = 130123, m = 11523 of the layer j = 3. The internal resource information cluster RM(130223, 29) belongs to the same layer j = 3 of the level h = 2. Business process diagrams of EPC II notation are simplified UML activity diagrams [6]. The meaning of symbols used in these diagrams is shown in the figure 9.3.

After the user's approval the graphic tool (fig. 9.1) writes down the structure of a given executive process to the tables SJA, RM, URSJA, YRSJA of the relational database, in which URSJA, YRSJA are tables of inputs (s, j, a, m, r) URSJA and outputs (s, j, a, m, r) YRSJA of located resources (m, r) RM to and from functional activities (s, j, a) SJA. Then the graphic subsystem automatically supplements the relational model and the graphic picture of the reference functional structure of a given business process control system with allocated and working functional processes [9].

In the example of the figure 9.4 the coordination unit is the only control unit, because it performs also functions of allocation and scheduling orders. Therefore

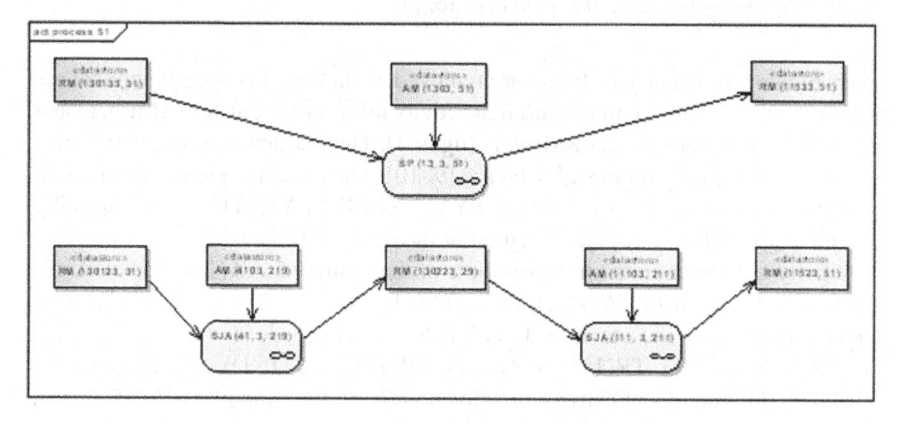

Fig. 10.2 EPC II activity diagram for a sample executive business process. Source [9]

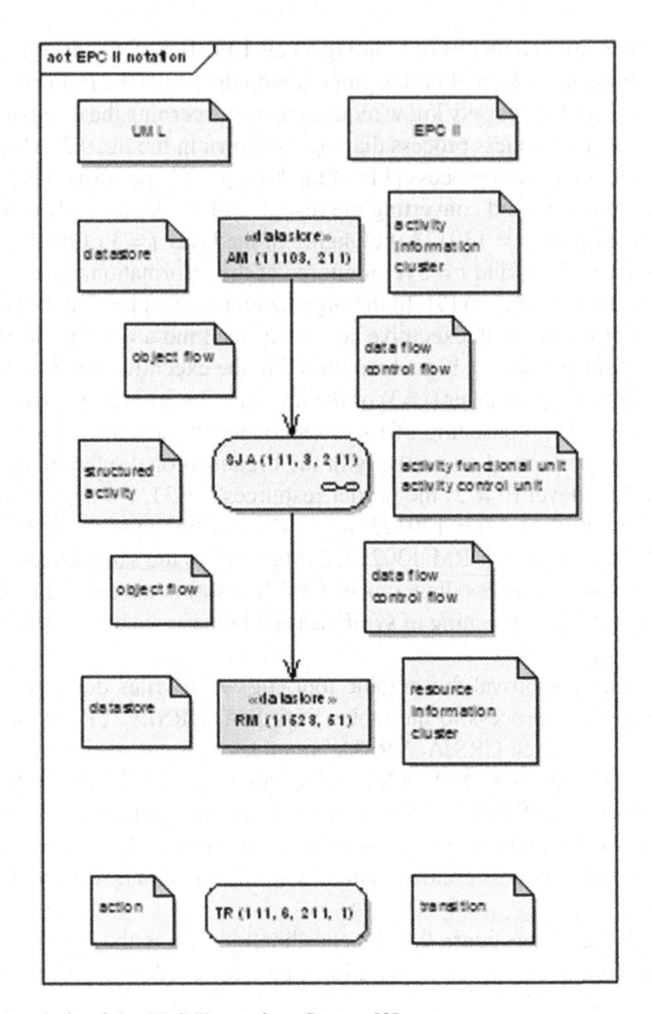

Fig. 10.3 Symbols of the EPC II notation. Source [9]

layers of allocation and scheduling are removed. Couplings between functional activities and their control units go through activity information clusters AM(m, a) and resource information clusters RM(m, r) (fig. 9.4). Their structure is stored in corresponding tables of the relational database [9, 10]. Their graphic picture is based on the tables SJA, RM, URSJA, YRSJA, SYAM, URSYC, YRSYC, whose meaning may be easily deduced from the diagram of the figure 9.4.

Activity transition units (in the example of the figure 9.5: TA(13, 6, 51), TA(13, 3, 51), TA(41, 6, 219), TA(41, 3, 219), TA(111, 6, 211), TA(111, 3, 211)) with single transitions (in the example: TR(13, 6, 51, 1), TR(13, 3, 51, 1), TR(41, 6, 219, 1), TR(41, 3, 219, 1), TR(111, 6, 211, 1), TR(111, 3, 211, 1)) are automatically created on the grounds of activity functional units (in the example: SJA(13, 3, 51), SJA(41, 3, 219), SJA(111, 3, 211)). Possible additional transitions, as well transition

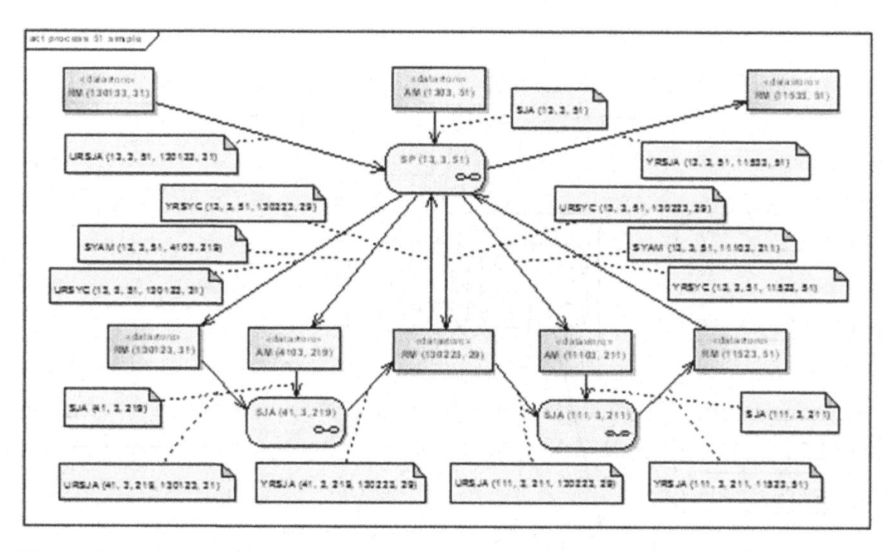

Fig. 10.4 An activity diagram for a simple control system of an exemplary business process. Source [9]

procedures must be introduced by users. Selection of procedures from the library of an EPC II system to its individual transitions is also the users duty.

10.3 Cooperative EPC II Systems

Building an EPC II system from scratch may be considerably sped up owing to the use of the framework EPC II system. Anyway, it may take a lot of time, even if existing software elements are used. Therefore it is worth to discuss application of the framework EPC II system to integration of information systems existing in the enterprise, or their parts.

In such cooperative EPC II system the framework EPC II system is only a kernel, whereas data are stored in the database, which already exists in the user's enterprise, while instead of the own library of transition procedures one can use procedures from an external repository of Web Services [1] (fig. 9.6) or from the user's database system (fig. 9.7). In the second case transitions of the framework EPC II system call existing user's procedures but do not transfer information between the procedures and the database (fig. 9.7). In both cases the framework EPC II system has to "know" addresses of transition procedures and identifiers of all attributes of records in the user's database. This condition is represented by 1 to 1 association between the list of transition procedures and the set of external procedures or the set of user's procedures, as well as between the set of attributes of information elements and the set of fields of rows in tables of the user's database. The contemporary information technologies, which enable fulfilling this conditions, are discussed in [1].

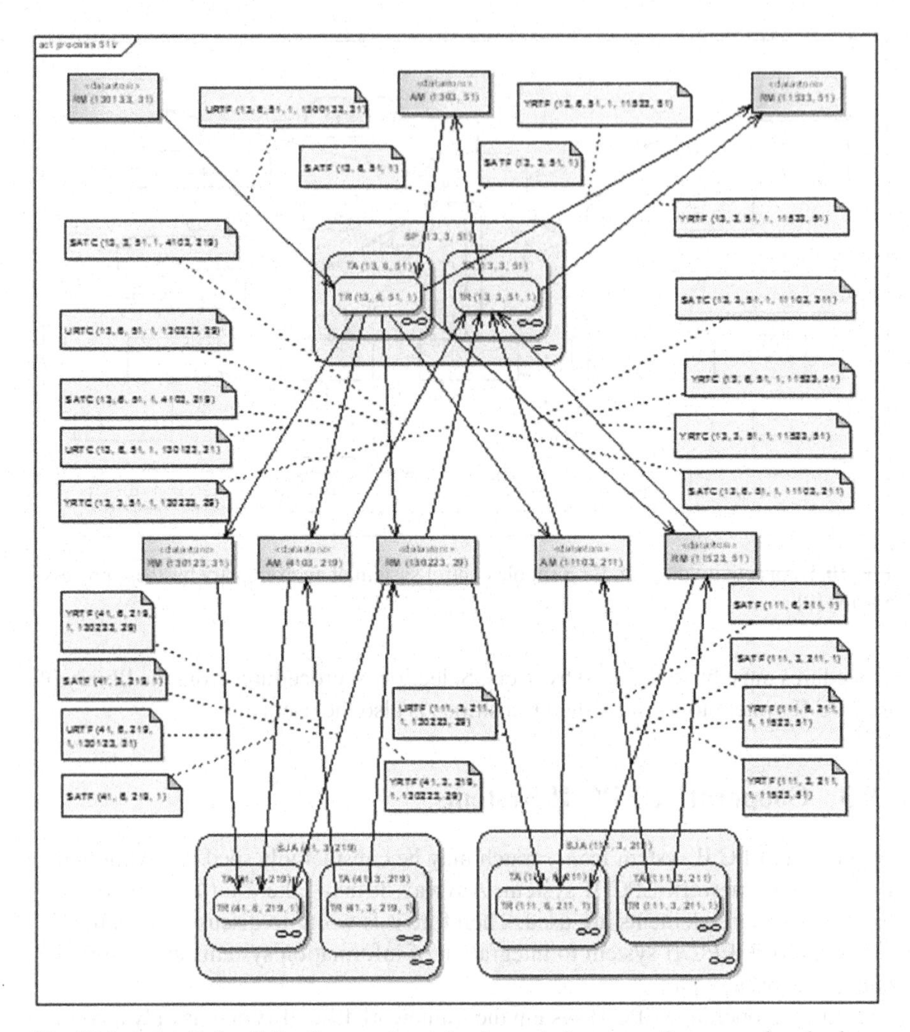

Fig. 10.5 A detailed activity diagram for a simple control system of an exemplary business process. Source [9]

10.4 The Framework EPC II System as a Simulator of Business Process Control

The framework EPC II system may work as a simulator of business process control systems. The structure of information [10,11] is the same as for the ordinary EPC II system, but in this structure only those system elements are active, which are objects of a given simulation experiment. For comparing effectiveness of different decision making methods or different management systems one can perform several simulation experiments with the same initial state and the same outside influences on the system but with different choice of active couplings in the reference coupling

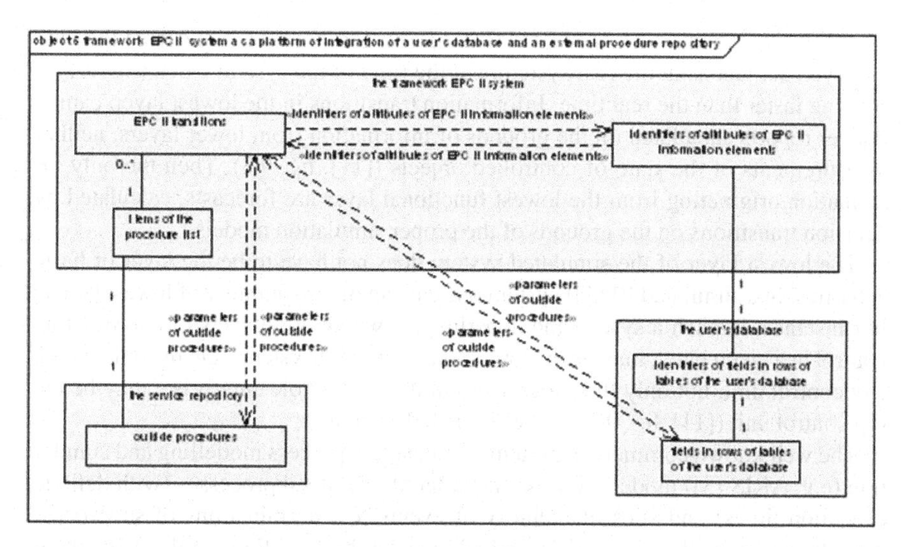

Fig. 10.6 Framework EPC II system as a platform of integration of a user's database and a service repository

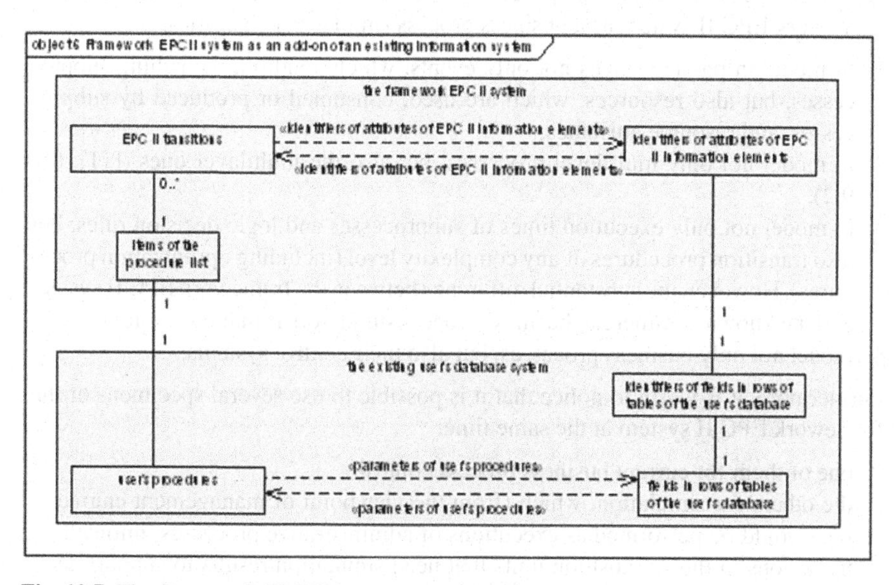

Fig. 10.7 The framework EPC II system as an add-on of an existing information system

structure and with different choice of transition procedures allocated to active transition units. Such comparative simulation experiments may concern the existing enterprise management system and the systems offered by different suppliers of ERP [5] software. On the grounds of such experiments, performed with various series of external influences, one can decide on the choice of the ERP software and on the purchase advisability.

All transitions in a simulated EPC II system, except for transitions from the lowest layer, act just as in the real system, but the time of the system clock (fig. 9.1) is running faster than the real time. Information transitions in the lowest layer, cannot deliver reports calculated on the grounds of information from lower layers, neither measurements of the state of controlled objects ([11], fig. 9.3). Then the only information originating from the lowest functional layer are forecasts, calculated by decision transitions on the grounds of the proper simulation models.

The lowest layer of the simulated system does not have to be the layer of basic activities. In a simulated EPC II system one can cut off any number of lowest layers. For instance, in such a system the departments, which are objects of coordinating control in a given plant, may be the lowest functional subsystems. On the other hand, any control unit, not only the supervisory unit of the whole enterprise, may be the top control unit ([11], fig. 9.1) of the simulated system.

The well known commercial systems for business process modelling and simulation (e.g. ARIS [3]) model processes as ordered sets of subprocesses (with definite execution times) and sites of memory of events (e.g. terminations of subprocess executions), as well as logic gates, which control the workflow on the grounds of information on events [2,3,4]. What is more, it is possible to model multilevel processes (whose subprocesses are processes of a lower level). The new features of the framework EPC II system as a business process simulator are abilities:

1. to model in passive stages not only events, which condition launching subprocesses, but also resources, which are used, consumed or produced by subprocesses, and resource attributes,
2. to model not only multilevel processes, but also the multilayer ones ([11], fig. 9.2),
3. to model not only execution times of subprocesses and logic decision rules, but also transition procedures of any complexity level (including optimization procedures). However, the substantial difference between the framework EPC II system and the known commercial business process simulators is that this system is able
4. model not only business processes, but also their control systems.

Furthermore, it is worth to notice that it is possible to use several specimens of the framework EPC II system at the same time:

1. one of them for current business process control
2. the others for simulation, which (from the viewpoint of management engineering) would be performed as executions of administrative processes, initiated by transitions in those transition units that need simulation results to support decision making (e.g. for decisions concerning business process reengineering).

10.5 Conclusions

The framework EPC II system may be applied in an enterprise of any trade and size

- as a skeleton of a new system of integrated management and process control,
- as a kernel of an EPC II system comprising also the existing database and servers of Web Services,

- as an add-on of the existing enterprise management system for reducing problems with adapting software to changes of business process structures,
- as a tool of modeling and simulating business processes and their control systems, which enables the more detailed analysis than in the case of currently available software.

In the case of the first three of mentioned above applications the framework EPC II system is an integration platform for all organizational levels and for all functional layers of a given enterprise, from the supervisory control unit to elementary processes inside workstations. This quality eliminates all structural problems of interfaces between systems of ERP, MES and SCADA-PLC. What is more, the alterations of business process structures, easily introduced to the system with the use of the graphic tool belonging to the framework EPC II system, cause automatic modifications of the whole system software. So corresponding human interferences in the enterprise software may be radically reduced.

On the other hand the data structure of the framework EPC II system is much simpler than structures of contemporary integrated management and process control systems. In the current state of the EPC II theory there are only around 150 information kinds (which are tables designed for storing information needed for management and control [10]). Moreover, there are only 26 structural attributes, which can be key attributes of information kinds. These numbers are quite small in comparison with thousands of tables in well known commercial ERP systems. What is more, EPC II systems comprise also MES and SCADA-PLC systems. Moreover, enterprises and their information systems are extremely diverse, whereas the framework EPC II system is the universal one.

All mentioned here reasons justify usefulness of the framework EPC II system as well as advisability of starting work on its prototype.

References

1. Buchwald, P.: Methods of integration of business logic and data in cooperative systems of enterprise process control. In: Knosala, R. (ed.) Computer Integrated Management, vol. 1, pp. 210–221. Publishing House of Polish Production Management Society, Opole (2011) (in Polish)
2. Business Process Model and Notation (BPMN), v.2.0 (2011), `http://www.omg.org/spec/BPMN/2.0`
3. Davis, R., Brabander, E.: ARIS Design Platform. Getting Started with BPM. Springer (2007)
4. ter Hofstede, A.H.M., van der Aalst, W.M.P., Adams, M., Russell, N. (eds.): Modern Business Process Automation: YAWL and its Support Environment
5. Ptak, C.A., Schragenheim, E.: ERP Tools, Techniques and Applications for Integrating the Supply Chain. The St. Lucie Press (2004)
6. Using UML Part Two - Behavioral Modeling Diagrams. Sparx Systems (2007), `http://www.sparxsystems.com`

7. Zaborowski, M.: The EPC theory. Basic Notions of Enterprise Process Control, Management and Production Engineering Review 1(3), 75–96 (2010), http://www.review.univtech.eu/images/archiwum/nr3/8-zaborowski.pdf
8. Zaborowski, M.: The EPC theory. Couplings Between Transitions in Enterprise Process Control Systems, Management and Production Engineering Review 1(4), 77–92 (2010), http://www.review.univtech.eu/images/archiwum/nr4/9-zaborowski.pdf
9. Zaborowski, M.: The EPC II theory. The Structure of Business Process Control Systems, Management and Production Engineering Review 2(1), 61–77 (2011), http://www.review.univtech.eu/images/archiwum/2011/nr1/9-zaborowski.pdf
10. Zaborowski, M.: The EPC II theory. Data Structure in Enterprise Process Control Systems, Management and Production Engineering Review 2(2), 53–70 (2011), http://www.review.univtech.eu/images/archiwum/2011/nr2/7-zaborowski.pdf
11. Zaborowski, M.: General Description of the Theory of Enterprise Process Control. AISC, vol. xx, pp. xxx-xxx. Springer, Berlin (2011)

Chapter 11
Selected Aspects of the Implementation Management in Multi-domain System Environment

Wojciech Filipowski and Marcin Caban

Abstract. Providing a change in an IT system, especially within the multi-domain environment of a large enterprise, requires proper planning and implementation of these plans in close cooperation and involvement of all stakeholders. Also problem managing and solving, in a range of aspects, is an important part of such considerations. Here the authors intend to bring these matters closer, to give an overview of most important issues and ways to overcome them showing variety of techniques appropriate to the situation/case.

11.1 Organization of the Project Team

In large projects, or projects conducted in large companies and for them, the project team organization is almost as important as planning of the scope of the change planning. The project structure needs to be effective in many aspects of project management to be able to provide best values from the client's point of view, while the client in this case is the company [1]. Lest us now take a look at some important matters of both: project structure and its values to the change implemented.

Central team. Although the name can vary from project to project, the - so called - central team is a part of the project structure that concentrates all aspects of management in order to prepare consistent analysis, reports and to help project management in having all key project parameters updated and accurate. The central team works in close cooperation with domain teams, which, as far as large enterprise is concerned, specialize in delivery management within their respective domains of responsibility. The term: *domain* is often understood as *system* and this is truth as long as the project operates on a integrated environment. Otherwise *domain* can refer different

Wojciech Filipowski · Marcin Caban
Silesian University of Technology
Faculty of Automatic Control, Electronics and Computer Science
ul. Akademicka 16, 44-100 Gliwice
e-mail: Wojciech.Filipowski@polsl.pl

A. Kapczynski et al. (Eds.): Internet - Technical Develop. & Appli. 2, AISC 118, pp. 131–140.
springerlink.com

domains of responsibility for certain tasks or task categories identified and assigned to teams by project manager (risk management, schedule management, resources, cash control and so forth).

Domain team. The domain team is part of the project structure that collects mainly substantive and partially management competence in a single domain (system). According to responsibilities identified by the project management and assigned to, the domain team, focuses on the quite a wide range of tasks - from: system analysis, delivery management, testing, to: vendor management, and procurement support. The domain team reports to the project via the Central team.

Subject Matter Experts. This is a special category of project participants. Subject mater experts are usually not strictly assigned to any part in the project structure, but since their unique competencies in certain key subjects, they are available directly to project management or consult both domain and central teams. Such kind of project resources are mainly extremely skilled and experienced specialists, who are hired by the project to solve the toughest issues form the fields of: engineering, law, and whatever field is of the project interest and need. Worth to mention is that SME are most expensive project resources so it is crucial that the project management decide to involve them on a limited basis and for well explained reasons.

11.2 Planning - Problems with Synchronization of Plans among Different Domains versus the Schedule Baseline

To describe this section let us imagine a project implementing a change in an integrated corporate environment. To simplify the description, we assume, that it consists of four phases: P1-analysis, P2-design, P3-testing, P4-production start-up, and there are three domains (systems) involved in the change, e.g. S1, S2, S3.

Another assumption is that the project baseline was defined earlier by the key stakeholders and then was given to the project management to confirm the change feasibility under defined conditions. The project role is to confirm the delivery plan, but in order to accomplish the task, there is a confirmation from all domains needed. There are different scenarios possible here, depending on the project objectives, parameters that drive the project, and the feasibility confirmation received from the domains.

Suppose, that the feasibility and schedule confirmations received from the domains are as presented on figure 10.1.

The figure shows how differences in confirmation between domains can influence the combined project plan.

To simplify the further deliberation let us draw a bottom line at the the following:

- the S1 domain vendor's confirmation showed that he was able to deliver the solution ahead of time;
- the S2 domain vendor confirmed his obligation as "just on time";
- the S3 domain vendor was not able to fit his plan into the baseline, so he confirmed with a delay.

Domain	P1	P2	P3	P4	Summary
S1	ahead	on time	on time	ahead	**ahead**
S2	on time	on time	on time	on time	**on time**
S3	ahead	delayed	delayed	delayed	**delayed**
Combined	**OK**	**NOK**	**NOK**	**NOK**	**DELAYED**

Fig. 11.1 Feasibility-schedule confirmation from specific system domains for each project phase

Each bid separately results in specific consequences, but the situation is much more complicated when it comes to combining these in one consistent project plan to deliver the integrated solution. From that point of view, not only delayed plans are problematic, but so are the ones too much ahead of time. Such differences need to put extra effort and resources to fix combined project plan so all the partial domain plans are interrelated to give the project schedule the acceptable level of compliance. There are several ways to achieve this goal [2]. We will now take a closer look at most common ones.

Now let us go through typical scenarios depending on the project drivers, meaning the main project constraints, which determine the factor that is unavailable for the project to change.

Scenario 1 - the project is scope driven (driver = scope)
When - for some reason - the project scope is fixed, or precisely: the project main foundation is to deliver the scope of change, there are ways to align the domain baselines by putting more resources and on the same ground more funds, to change the combined project plan, to achieve the goal of saving the scope.

Scenario 2 - the project is schedule driven (driver = time)
This variant is used when the main project constraint is time, meaning the baseline plan cannot change. The time constraint is one of most serious ones projects are to deal with, because to work out the consistent plan, apart from putting more funds and resources, there might be a need to reduce the scope, which for each project is a critical decision.

Scenario 3 - the project is cost driven (driver = $)
When funds are the constraint factor, the project is really hard to manage, because each decision made by the project management can be noticed from the cost criteria viewpoint. Especially when the project operates on the complex integrated environment of a large enterprise, having limited amounts of money to run, it is at the same time inspiring and restrictive. In such conditions to be able to deliver the change, all possible ideas are acceptable as long as we can receive approval from the project stakeholders. In this case one can expect that the aligned and combined baseline is an outcome of change in planning as well as scoping and risk assessment in all domains taking part in the project.

As the experience shows, the ways that best solve the project planning issues are strongly depended on the project driver and confirmations received from the domains. If the project has such an opportunity under the real conditions to use one of the propose methods, than it is a situation of most effective problem solving. However in many real situations project do not have the comfort of replanning before they start, or the issues come up when the project is already in progress. This part of project life is addressed to in the next section of the article, that covers the change management topic.

11.3 Managing Unplanned Changes during the Implementation Process

There are some situations during the project course, when the scope of work needs to change. It happens rather more often in larger projects, but since the variety of reasons, each - even a single domain - project needs to be prepared for them. These unplanned events are called change requests. Again, no matter what name is used to address a change in the project - each time such a thing occurs, it needs to be analyzed, planned, designed, tested and applied. This article does not treat about elementary implementation phases mentioned above, we want to focus more on the reasons of the change occurrence, instead, and how it can be handled form the project point of view.

As for the reasons - change requests can be triggered by different factors, including: analysis fault, change of business requirements, technical issues discovered during the design phase, efficiency issues discovered during the test phase, additional services needed - additional system environment installation/configuration, extended test support, legislation, etc.

Since change requests are often costly, as well as time and resource consuming, they need to be carefully handled in the project. Each project really needs to develop and apply the change management procedure to follow a consistent and orderly way of change request implementation [3].

On Figure 10.3 is an exemplary change request form template in the form of extracted selected sections of it, so one can have a vision of how important it is to manage such unplanned changes, and to take the key project factors into account while processing a change request, to be able to continue the project without a major shock.

11.4 Test Management

Testing plays quite an important role in the process of application a change to an IT system. The success factors here, are:

- to chose a proper type of test/tests to the unique situation taking into account: all the possible aspects of the change, the system and the specificity of an enterprise - the client;
- to perform the chosen tests as planned with the agreed level of services when it comes to delivery of bug fixes;

Change Request name	*To be filled out*
Change Request Number	*To be filled out*
Project Name	*To be filled out*
Project Id	*To be filled out*
Project Business Owner	*To be filled out*
Project PMI	*To be filled out*
Required delivery date	*To be filled out*
Guaranteed delivery date	*To be filled out*
Change Request reason	*To be filled out*
Change Request responsible stakeholder	*To be filled out*

Fig. 11.2 Change request identification section

- to manage the testing effectively during the project course so the process could be seen not only as time consuming and costly event, but mostly as effort to improve of the quality of implemented solution.

While the firs two bullets are covered in another article, the third one will now be briefly taken care of.

Software testing is a key factor towards its reliability an other parameters, as: accessibility, maintenance cost, especially in branches where the quality of the solution provided to the end user is crucial.

According to complexity and specificity of a project, different classes of tests are considered, including:

- Unit Tests;
- System Tests;
- Version Upgrade Tests;
- Link Tests;
- Integration Tests;
- Regression Tests;
- Efficiency Tests;
- Trial-run Tests;
- Client acceptance tests.

Business requirement description
Please fill in the structured and numbered manner *R1.* *R2.* *R3.* ...
Business process diagram
To be filled in
General description of the technical solution
To be filled in
Detailed description of changes in individual systems (domains)
Domain 1. *Domain 2.* *Domain 3.* ....
Change request delivery plan
Domain 1. *Domain 2.* *Domain 3.* ....
Risk management
Risk identification *Risk analysis* *Risk management plan (scope, cost, influence on the project plan)*
Change request cost
Domain 1. *Domain 2.* *Domain 3.* ....

Fig. 11.3 Change request template (an example)

To perform all chosen types of tests, project management needs to make a good planning beforehand, and then, while the required tests are run, to assure an effective testing management [4]. There are tools to help in managing tests in projects. Those are called bug trackers, but these days functionality of such tools goes far beyond fault recording. They enable to:

- store test scenarios (test case sheets) with decomposition to single test steps;
- assign testers to each test case;
- track the both progress and status of each test;
- report bugs;
- manage bug analysis;
- report bug fix design status;
- calculate bug fix timing;
- prepare reports;
- communicate between test process participants;
- archive the communication and other activities.

11.5 Quality Parameters (SLA)

This section explains how can service quality be enforced in the project, while services are activities performed by project members or hired personnel - including vendor - additionally to the contracted product delivery. The scope of those services is defined either in the project work-breakdown-structure or in a contract with vendor (or both), depending on who is responsible for providing certain services, while the level of quality of services is determined and agreed in a document called: Service Level Agreement (SLA) [5]. This document is either a part of a contract with a vendor or a piece of the project documentation. Below is an example of a part of an SLA agreed in a project of implementation the change to the complex integrated environment of a large enterprise. This is about services provided in testing, and involve support for testing, analysis, and bug fixes delivery.

Below is an explanation of most important SLA nomenclature:

- hours (h) and days (D) should be understood as respectively: clock hours and days;
- Response Time - the time that runs from the ticket issue - an action performed in the form determined by the SLA contract parties;
- Time Window - Time of support served for tickets issued;
- Workaround - an initial stage of bug-fix process that allows to use the functionality of the System until the target repair (bug fix) is delivered;
- Bug fix - a permanent restoration of System functionality and performance
- Ticket classification:
  - A-category: the System inability to meet the key business process, or a failure to meet key performance requirements for the System;
  - B-category: an unavailability or defect in the key system functionality or a serious error in the operation of business process;
  - C-category: an unavailability or a defect in the operation of system functionality or business process.

Factor	Ticket category	Level of service	Level of service	Level of service
Level of support for ticket solving		**GOLD**	**SILVER**	**BRONZE**
Response Time [h]				
	A-category	2 h	3 h	4 h
	B-category	3 h	8 h	10 h
	C-category	6 h	12 h	18 h
Required Level [%] (tickets served on time/all tickets)		90%	90%	90%
Business service restore time [h]				
	A-category	8 h	12 h	24 h
	B-category	16 h	24 h	48 h
	C-category	120 h	120 h	120 h
Bug fix delivery for a ticket [d]				
	A-category	2 D		
	B-category	5 D		
	C-category	7 D		
Required Level of Bug-fixes delivered on time per all tickets [%]		100%	95%	90%
Application availability		**GOLD**	**SILVER**	**BRONZE**
Required Level of availability of the application In month [%]		99,5%	98,0%	95,0%
Time window		**24/7**	**12/7**	**8/5**
Required service window		24 h/D 7 D/week	24 h/D 7 D/week	8 working h/D 5 working D/week

Fig. 11.4 A set of SLA parameters (an example)

On Figure 10.5 are displayed general dependencies between the response time and bug fix delivery.

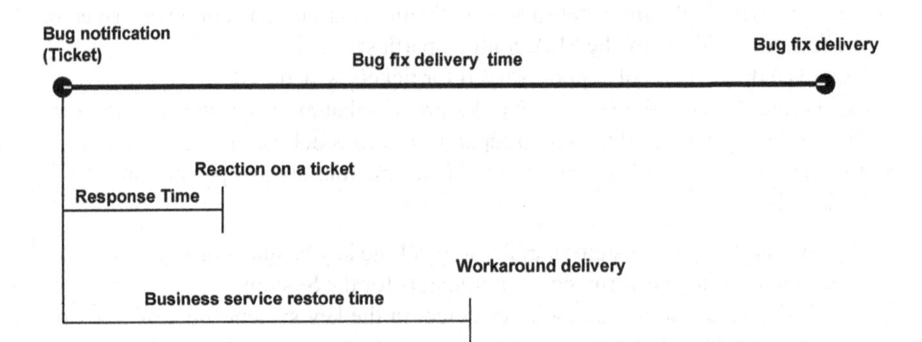

Fig. 11.5 Time dependencies between the activities in the bug fixing process

11.6 Vendor Management (Contractual Issues - Costs, Penalties, Success Fee)

This section of the article will briefly go through selected aspects of vendor management in large scale projects, as such projects tend to use subcontractors to perform virtually any activity from management support to specialized engineering tasks and consulting [6]. Each relation with a subcontractor no matter if it is an analyst who performs a single task or a vendor which runs a project stream or delivers software, needs to be formalized in the contract form [7].

In each contract there are certain elements that secure both parties, including: Price, penalties, success fee. All of these contract sections are related to money, but each of them in different aspect. Price is an overall cost of the contract that is an element of often hard negotiations between parties.

Penalties are means of compensate the loss for the client and simultaneously punish an unreliable contractor. This is an ultimate resort when things go very wrong, and form the commercial point of view this is an end of the cooperation for the parties for quite a long time.

Success fee - this is not as popular as one would think but a very kind and useful mean to summarize the fruitful cooperation, especially when parties intend to continue working together in the future. It is an extra reward for a reliable supplier.

11.7 Conclusions

While single domain projects or projects of a relatively smaller scale can be conducted with a relative ease, the multi domain or large scale ones need to be organized from the very beginning with attention and understanding of complexity of the matter. The authors hope that this article could help to put a little bit of light on some of the matters of significant importance on how important it is for each large scale project to be sensibly organized and conducted.

References

1. Englund, R.L., Graham, R., Dinsmore, P.C.: Creating the Project Office: A Manager's Guide to Leading Organizational Change (February 2003) ISBN:978-0-7879-6398-9
2. Agoulmine, N., Dragan, D., Gringel, T., Hall, J., Rosa, E., Tschichholz, M.: Trouble Management for Multimedia Services in Multi-Provider Environments. Journal of Network and Systems Management 8(1) (March 2000)
3. Hiatt, J., Creasey, T.: Change Management; The People Side Of Change (September 2003) ISBN:978-1930885189
4. Baresi, L., Pezze, M.: An Introduction to Software Testing. Electronic Notes in Theoretical Computer Science 148(1) (2006)

5. Scheikowski, A.: Service Level Agreement Best Practices - Templates, Documents and Examples of SLA's in the Public Domain PLUS access to content. theartofservice.com for downloading (April 2010) ISBN:978-1742443065
6. Saxena, A.S.: Enterprise Contract Management: A Practical Guide to Successfully Implementing an ECM Solution (March 2008) ISBN:978-1932159905
7. Emmett, S., Crocker, B.: Excellence in Supplier Management. Cambridge Academic (March 2009) ISBN:978-1903499467

Chapter 12
Electronic Recruitment System for University Applicants as B2C and B2P System

Paulina Puścian-Sobieska

Abstract. The author's deliberations in the article deal with the possibility of classifying the Electronic Recruitment System for university applicants as B2C as well as B2P system.

12.1 Introduction

The decision about creating the electronic recruitment system for university applicants was made at the University of Łódź in 2003 [11]. Since at that time the access to the Internet in Poland varied between different parts of the country — that is, in some areas (mainly in cities) Internet access was common while in towns it used to be impeded, it was decided that the pilot version of the system would be created and in 2003 both „paper" and electronic recruitment were carried out. The pilot version was such a big success (more than 35 percent of applicants had chosen the electronic option) that the next year and of course over the following years too only this way of recruitment was practiced [12]. Finally, Electronic Recruitment System (ERS) became as it were a part of Vice-chancellor Office System which supports various areas of state university central administration [7, 8, 13].

In the natural way, technological evolution was accompanied by the evolution of consciousness of both the users of ERS and decision-makers, concerning what one can expect from this system [13]. According to the principle „The appetite grows with what it feeds on", people who were factually connected with the recruitment process asked for additional functions which could make their work easier and streamline the registration process and, generally, the recruitment. Eventually, ERS, at first being the system supporting a candidate, became a powerful and complex system of circulation and management of documents [5]. It began to be noticed that such system was becoming one of the most important ways to contact with

Paulina Puścian-Sobieska
Informatic Center, University of Łodź, Lindleya 3 st., 90-131 Łódź, Poland
e-mail: paulina@uni.lodz.pl

A. Kapczynski et al. (Eds.): Internet - Technical Develop. & Appli. 2, AISC 118, pp. 141–146.
springerlink.com

a potential applicant for studies even before such person would like to start the registration process. The candidate will find the system and learn about the whole teaching offer and their possibilities. Nowadays, when there are about 300 majors or specializations in the offer of University of Łódź, ERS is the basic source of information for applicants for studies. Of course, paper version of compendiums of university courses is still published, however, they are treated as the form of a leaflet which is to encourage people to become interested in a given university's offer.

It is obvious that Electronic Recruitment System changed through the years [12]. What is more important is the fact that those changes were not connected with the technological aspect (which basically remains the same) but so-called business logic or the philosophy of action. Interestingly, those changes did not result from flawed initial premises in any way but they were the result of simplifications, which were made consciously to make the system mature together with the users. It was a deliberate design decision to arrange a big system of information circulation with the use of a software life cycle based on incremental model[9] or incremental development [10].

The author is not familiar with any research that analyzes either successes or failures of similar systems in the world. However, the literature provides an analysis of the online selling systems [4] in which it has been noticed that transferring sale to the internet shops attracts consumers' attention and increases sales outcomes. On the other hand, the work [3] deals with the problem of trust while selling financial services. Finally, the research [2] discusses issues concerning profits, adversities and crucial factors for the success of B2B undertakings. Thus, in this work, the author is trying to make the analysis of the characteristics of Electronic Recruitment System from the consumer's point of view (an applicant for studies) and to make the classification based on the definitions of terms used in economics and information science, such as B2B, B2C, e-Marketing and e-Commerce.

12.2 Computer Systems from Consumer Point If View

Let's start from the classification of computer (electronic) systems from the consumer point if view. The notion of computer systems which support commerce, known also as electronic commerce systems (e-commerce systems) shall be understood as those which support the sale of goods and services, taking and confirming orders and handling of cashless payments with the use of electronic channels. Usually, one can also distinguish m-commerce which uses mobile computer devices, however, in this work, we will equate these two terms assuming that they differ only in technological aspect (communication channel). Let's go back for a moment to the very term of „electronic commerce" and following [1] introduce the following typology:

1. **Direct e-commerce** — where the whole commercial transaction — as from finding the offer in the Internet, through the process of formulating and placing an order to its realization and payment — is carried out only with use of electronic channel of communication which is the Internet.

2. **Indirect e-commerce** — looking for commodities, services, new contractors, sending documents and payments- all take place via the Internet, whereas the delivery of an ordered good or service is in direct and traditional way.
3. **Hybrid e-commerce** — other intermediate forms are temporarily used because of inadequacies of the Internet and telecommunications.

If one takes into consideration contracting party, taking part in the communication process, the following kinds of commerce can be distinguished [6]:

B2B (business-to-business) — electronic commerce in which orders and sales transactions are exclusively between businessmen.
B2C (business-to-consumer) — electronic commerce between an enterprise and a consumer (mainly retail sales).
B2P (business-to-public) — electronic commerce in which transactions are conducted between an enterprise and the public sphere (mainly the exchange of information which is essential for the functioning of businesses and their potential customers).
C2P (consumer to public) — transactions between consumers and public infrastructure media, the shift of information exchange from the public sphere and an institutional client to an individual client.
C2B (consumer-to-business) — transactions between a consumer and a company which are not directly connected with the course of commercial transactions (so-called the group of factors accompanying commercial transactions such as creating brand image, building up trust in company, image, logo, name, the promotion of goods and services).

Aside from areas strictly equated with electronic commerce the communication between:

C2C (consumer-to-consumer) in which there is an exchange of information and even services between individual members of society is also possible.

12.3 Classification of Electronic Recruitment System

Bearing in mind the division presented in the previous part we can now try to classify the ERS system. While analyzing ERS users — with a state university (business) on the one hand and an individual client on the other — it can be easily found that ERS is B2C system. Since the ERS is a system which is supposed to win over a potential purchaser of the services that are offered by a state university, while building it a particular attention was given to a user's suitable interface and easiness while using the service. What is the main function of ERS is that a candidate chooses a major among those offered by the University of Łódź (the registration of a given person as a potential candidate), so similarly to B2C systems which focus on selling goods or services to potential purchasers.

From the short analysis of ERS the following functions characteristic to B2C system appear:

- ERS clearly presents the catalogue of its products as from a simple major search engine (by entering the name of a major or a given phrase) through the possibility of searching in a hierarchical tree. Each product — a major — has a detailed description and the rules of recruitment on the separate sub-website.
- ERS website has an elaborate client identification system. One can use the service after previous registration and establishing a login and password. The system prevents the situation of having several accounts by one user because personal identity number (in Polish citizens' identity cards and in the general register office) is an account identifier which is checked by the application if it is in keeping with the date of birth and user's sex. Data on candidates (clients) is stored in a private database in order to protect it against potential online theft.
- Electronic Recruitment System cooperates with other computer systems and commercial applications, for instance documents processed by the system are saved as .pdf, .xls, .doc formats.
- ERS has an extremely user-friendly and clear interface which can be used in many search engines. Interface is created in a graphic mode which is easy to operate and pleasing to the eye. The service contains all the important information needed for the correct recruitment process. The process has several stages (data which must be entered is grouped thematically and modularly). Forms are clear and lucid and, in most cases, verify the entered text with respect to its compliance with data format. It is a very important part of B2C system which enables the correct use of the website by organization's clients.
- The main advantage of moving the recruitment into the virtual space in the Internet is having the access to the system from every place in the world (having the Internet signal) which is one of the main characteristics of B2C.
- The next important feature which lets us classify ERS as B2C system is the complexity of the service. The system enables browsing and comparing products that are offered (majors); it enables choosing a major (there are no restrictions concerning the number of majors that one can choose, one cannot only recruit for one major several times). The system allows completing all formalities that are essential to enter the recruitment process including the management of a „purse" with money paid into organization's account (the accounting of money at the university bank account is visible immediately in a candidate's private account). ERS is also a channel of communication between the Instruction Department (responsible for correct recruitment), secretaries of individual majors (supervising correct recruitment on a particular major) and an individual candidate.

The characteristics of B2C system can be observed while analyzing the history of ERS development:

1. At first the system offered only the survey of majors that were offered and the rules of recruitment so it was an ordinary website. In its most basic form it was only giving the instructions on how one should recruit but it did not allow it.
2. After the extension of the system in the next year the system allowed recruitment through the Internet as well as in the normal so-called „paper" way, connecting the virtual way with the paper one. The extension enabled clients to choose a

service and to follow its realization (recruitment process) online without any need for assistance from the staff.

3. The next step was providing clients with direct assistance in the recruitment process (an Internet café with trained staff, technical assistance via emails and phone support, sms notices) thanks to which clients could get immediate answers to their questions and remove doubts which could previously hinder the choice of a major in offer.
4. The next stage which makes the recruitment reach higher level of B2C system is creating individual bond with a candidate. It is planned to create a permanent account for each candidate via which it would be possible to have lasting CRM system (Customer Relationship Management). What is also planned is free email on the university server and generating adapted services which would reflect individual client's preferences connected with the choice of one's major.

The above analysis shows that ERS accomplishes the principles of B2C but is it only that? The system is also the medium used by the university to create its image as being modern, complying with the highest teaching standards and geared towards the fulfillment of customer needs. And these are the characteristics of B2P system which develops a company relations with the surroundings by means of Public Relation, market research and marketing. In order to prove the presented thesis several important elements of ERS should be listed as they accomplish the principles of B2P:

- Electronic Registration System has, apart from the user interface, elaborated administration which not only allows to keep statistics but also to make simulations which are aimed at eliminating the problems connected with the application functioning and easy planning of recruitment for the next year. Thanks to it, limits on students for particular majors can be modified what gives the opportunity to adjust the offer to customer needs.
- Documents which are processed and generated from ERS undergo full personalization. Decisions as well as notices that are sent address candidates individually (device that is used by a great number of commercial institutions such as banks and mobile communications etc.).
- It is possible to send short text messages from the website concerning the recruitment process or informing about deadlines and changes thanks to which a candidate can react to given information immediately (the same function is realized simultaneously via an email and an announcement that is placed in candidate account).
- ERS was presented at many scientific conferences and got recognition from many universities which is proven by the fact that the system was implemented in Police Academy in Szczytno in 2009 (the implementation was done by the employees of the University of Łódź Informatic Centre). The service has been used by the above mentioned university for three years.

12.4 Summary

ERS has got an opinion as being complex recruitment system (recruitment, information, document circulation, reporting). ERS was created totally by the employees and students of the University of Łódź which proves not only the teachers' great competence but also a solid and practical training process, which are a great mark of our university.

As a matter of fact, in 2010 it was decided to use an alternative solution at the University of Łódź. It resulted from the fact that the ERS did not provide for priorities while choosing a major. Nevertheless, the system is still successfully used at the Police Academy in Szczytno and the University of Łódź is considering to return to the system.

References

1. Chmielarz, W.: Systemy biznesu elektronicznego (Systems of the electronic business). Diffin, Warszawa (2007) (in polish)
2. Dubelaar, C., Sohal, A., Savic, V.: Benefits, impediments and critical success factors in B2C E-business adoption. Technovation 25, 1251–1262 (2005)
3. Kaplana, S.E., Nieschwietz, R.J.: A Web assurance services model of trust for B2C e-commerce. International Journal of Accounting Information Systems 4, 95–114 (2003)
4. Liaoa, Z., Cheung, M.T.: Internet-based e-shopping and consumer attitudes: an empirical study. Information & Management 38, 299–306 (2001)
5. Miodek, K., Sobieski, Ś.: Architektura i bezpieczeństwo systemu elektronicznej rekrutacji kandydatów na studia (Architecture and safety of the Electronic Recruitment System for university). In: Nowe Technologie Sieci Komputerowych, vol. 2, pp. 483–489. Wydawnictwa Komunikacji i Łączności, Warszawa (2006) (in polish)
6. Niedźwiedziński, M.: Materiały na VII Krajową Konferencję EDI-EC Electronic Data Interchange-Electronic Commerce, Łódź-Dobieszków (1999) (in polish)
7. Nowakowski, A., Sobieski, Ś.: Evolution of Information Systems at the University of Lodz during 2002-2007 — the Report of Achievements and Failures. In: Proceedings of Eunis 2007, Conference, June 26-29, University of Grnoble (2007)
8. Nowakowski, A., Sobieski, Ś.: Integrated Information System for University of Łódź. In: Proceedings of Eunis 2005 Conference, July 20-24, University of Manchester (2005)
9. Pressman, R.S.: Software Enginering — A Practitioner's Approach. European Adaptation, 4th edn. McGraw-Hill (2000)
10. Sommerville, I.: Software Engineering, 5th edn. Addision-Wesley Publishers (1996)
11. Sobieski, Ś.: Elektroniczna rekrutacja na studia wyższe (Electronic Recruitment System for university). In: Internet 2005, vol. I, pp. 261–271. Oficyna Wydawnicza Politechniki Wrocławskiej, Wrocław (2005) (in polish)
12. Sobieski, Ś.: Ewolucja systemu rekrutacji na Uniwersytecie Łódzkim w latach 2003-2005 (Evolution of the Electronic Recruitment System for University of Łódź in 2003-2005). In: Aplikacje Technik Multimedialnych w Organizacjach Gospodarczych, pp. 77–81. wydawnictwa zwarte Wyższej Szkoły Ekonomii i Administracjii w Kielcach (2006) (in polish)
13. Sobieski, Ś.: Koncepcja systemu informatycznego wspomagającego pracę państwowej uczelni wyższej oraz omówienie wybranych modułów (The project of the Electronic Recruitment System for university and the main module). In: Bazy Danych: Modele, Technologie, NarzĘdzia, vol. II, pp. 273–279. Wydawnictwo Komunikacji i Łączności, Warszawa (2005) (in polish)

Part III
Information Security and Business Continuity Management

Chapter 13
Computer Continuity Management in Distributed Systems of Management and Steering

Andrzej Grzywak and Piotr Pikiewicz

Abstract. The article presents the monitoring method of computer operation continuity in distributed systems of management and steering. The article concentrates only on factors connected with the continuity of computer systems operation. The method of monitoring and monitoring results of the system aiding the management of a chosen institution has been presented in this paper. The goals of this prepared method are the assessment of the system operation, improvement of servicing and managing as well as a collection statistic data that let build the system's mathematical model.

13.1 Introduction

Business Continuity Management (BCM) is the intentional process of establishing the strategy and organizational structure of a company whose aims are:

- active improvement of a company resistance to disruptions of its stability that would make impossible to realize the company's strategic goals,
- ensuring methods that let restore the organization's abilities to produce its key products and services on the appropriate level and in the appropriate time after the event occurrence,
- ensuring reliable tools and management methods in crisis situations and the protection of a company reputation and brand [2] [3].

In the article the research methodology of computer systems operation continuity has been presented as well as the original method of computer-aided management system monitoring. The reliability of the computer system can be assessed by the particular elements monitoring. The telecommunication systems reliability monitoring enables to detect the existing critical events, what is useful for the appropriate

Andrzej Grzywak · Piotr Pikiewicz
Academy of Business in Dabrowa Gornicza, Department of Computer Science
ul. Cieplaka 1c, 41-300 Dabrowa Gornicza, Poland
http://www.wsb.edu.pl

A. Kapczynski et al. (Eds.): Internet - Technical Develop. & Appli. 2, AISC 118, pp. 149–154.
springerlink.com

technical services, and it makes possible to conducts statistic research. On the basis of the research, the improvement of reliable parameters such as Mean Time Between Failures(MTBF) and Mean Time to Repair(MTTR) can be obtained by means of redundancy methods. The proposed research thesis is a question about the interrelation between the correct system aiding management functioning and the maintenance of a company computer system operation continuity including telecommunication systems. The basis to formulate this thesis is the fact that business processes occurring in a company, to a large extant, are realized with the application of computer techniques. It is presumed that the widely understood computer network and ensuring the operation of all network junctions are crucial elements for the company proper performance [1] [5] [11]. The devices of the telecommunication systems are arranged in a local network, called the Intranet or Local Area Network (LAN) which is connected with external users with the application of wide network tools (the Internet, Wide Area Network) enabling the exchange of information [6]. In the case of multi-department organizations, when these departments are situated in different geographical locations, the connections among them are realized by means of private networks on the basis of leased links in the framework of a public network or special channels - virtual private networks VPN. The compression or ciphering can happen in such channels in order to ensure better quality or better data security. The solutions based on VPN can be used in organizations whose users take advantage of data placed in a telecommunication system. The system is available from different geographical locations [10].

Possibly all elements of the system(Fig. 13.1) should be embraced by a monitoring detecting critical conditions. Local networks (Measurement Unit 1 and 2), connections among departments (VPN Measurement Unit) and services assigned for external users (Measurement Unit 3) should be monitored.

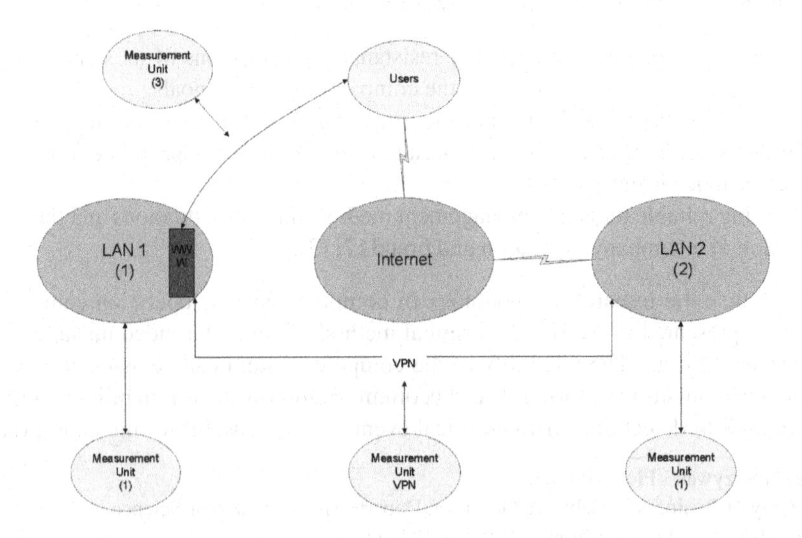

Fig. 13.1 WM type monitoring system

13.2 Concept of a Measurement Method

The applied method is based on agents' concepts that are system's chosen appliances [2]. The agents collect information which is transferred to a monitoring unit. The following tools were used in this method:

- **Windows Management Instrumentation (WMI).** – a set of protocols and the Windows extensions enabling management and access to a computer stock, for example, network adapters, currently open programs, a list of processes, data acquired from built sensors [13],
- **Simple Network Management Protocol (SNMP).** – a family of network protocols used to manage the network appliances such as routers, switches, computers or telephone switchboards [5].
- **Performance Counters (Windows).** – MS Windows counters used to provide information about operation performance of the system's applications and services.

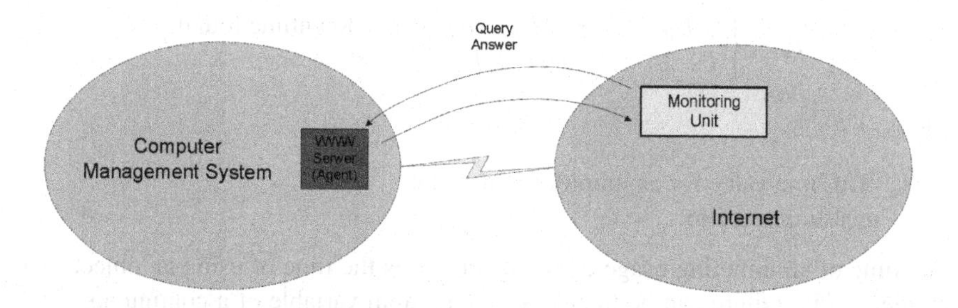

Fig. 13.2 WM type monitoring system solution (3)

In the Measurement Unit 3 (Fig. 13.1), the monitored element is the availability of WWW services (Fig. 13.2). In this case, the confirmation of service availability from a local network can only prove the correct performance of all external system elements that are responsible for delivering a service to external customers. Beyond these elements, the condition of customer's connection with the system influences the availability of the WWW service realized in a public network. For the lack of public network tools monitoring possibilities, the service availability can be monitored while simulating service user's operations. To this end, an application generating query to a service and checking the correctness of answers should be implemented to a monitoring unit. A data measuring and acquiring system can be arranged according to the centralized and distracted architecture. This division was achieved according to the place of storing the gained measuring data.

13.3 Method of Measuring Data Analysis

Data gained thanks to implemented measuring mechanisms must be processed in order to make the information about the event occurrence, the lack of operation continuity for all monitored appliances available. Information about the event is used to define elements of organization infrastructure reliability factors. These factors let assess reliability parameters of the whole system, select of appliances connections and propose the redundancy in such places where it is necessary and the most effective.

Fig. 13.3 presents the method of appliance crash detection on the basis of a measured UpTime parameter. On the basis of UpTime measured value, a discrete function of appliance operation discontinuity in time for every appliance is calculated. The applied measurement method enables to define the operation discontinuity time for successive critical events (Fig. 13.3) with the precision bigger than the period between successive detection points of UpTime in time parameters. The time value of a failure Tn is calculated per sample (1):

$$T_n = \begin{cases} U_n & \Leftrightarrow \quad U_n > U_{n-1} \text{ ,No downtime found} \\ T - U_n & \Leftrightarrow \quad U_n \leq U_{n-1} \end{cases}$$

where

U_n - UpTime value for an umpteenth measure,
T - measurement time.

The time of an unfailing usage of an object that is the time of using an object to a moment of its failure can be treated as a T random variable of a continuous type. Continuous functions, with respect to time such as: probability density f(t), fallibility F(t) (distribution function), reliability R(t), intensity of failures λ(t) are characteristic features of a T random variable. In an explicit way, each of these functions defines the T random variable determining, at the same time, a form of the rest of these functions. Fallibility that is the probability of object damage before a t moment is defined by a distribution function of a T random variable. The Weibull decomposition is often used in a reliability theory for mathematical modeling time of unfailing object exploitation distribution.

A selection of probability distribution requires gaining a big amount of information about critical events. Big reliability of modern computer systems causes that gaining a sufficient amount of data requires conducting a monitoring for a longer period of time.

13.4 Results of Conducted Measuring Research

Gaining measuring data from a conducted monitoring of telecommunication devices operation continuity and network servers connected to work stations was carried out in the period from May 2010 to September 2010. A big number of events

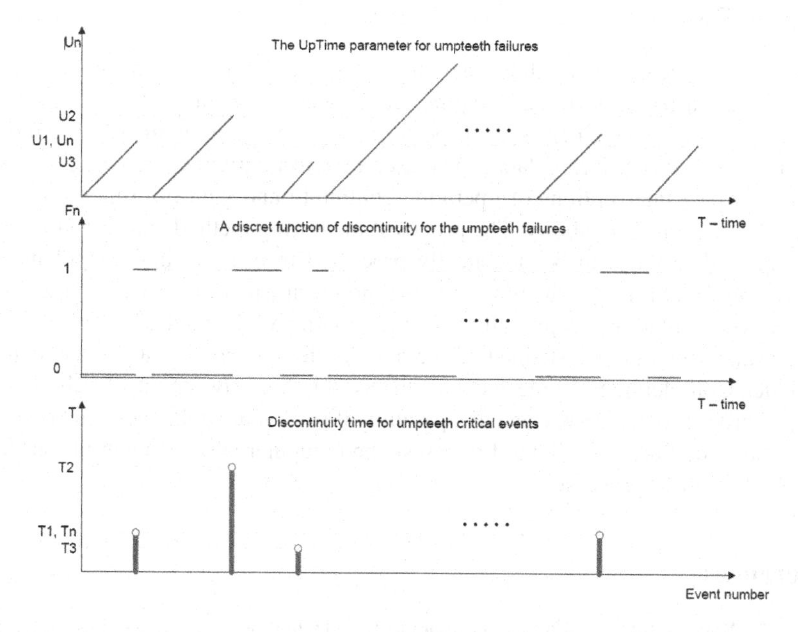

Fig. 13.3 Failure detection on the basis of the UpTime parameter

disrupting the operation continuity was noticed. The defined reliability factors values for network appliances that were used to maintain educational business processes are presented in Table 13.1.

Table 13.1 Reliability factors of network appliances

An appliance	Number of events recognized as critical	Defined MTTR time	Defined MTBF time
Junction 1	4	0,25 h	900 h
Junction 2	3	0,66 h	1200 h
Junction 3	5	0,42 h	720 h
Junction 4	7	0,20 h	514,28 h
Junction 5	6	0,25 h	600 h
Junction 6	8	0,30 h	450 h
Junction 7	8	0,27 h	450 h

For network appliances the reliability factors show better behavior of retaining operation continuity. Similar time of events discontinuity occurrence lets presume that they were caused by the same external element. The events analysis confirmed that the operation interruptions were caused by problems with appliances power supply.

13.5 Summary

The concept of measuring modules, with the application of mentioned mechanisms, implementation let achieve the computer appliances, necessary for the organization appropriate management system performance, operation continuity monitoring. The collected measuring data can be used to detect current appliances failures, reconstruct past interruptions of operation continuity and gain information about the lack of services availability. The proposed measuring method can serve to gain and acquire data that can be statistically processed in order to define appliances and services reliability parameters. A remote monitoring method lets also check the web services availability empirically what can be compared with these services predicted availability that was defined on the basis of the system's created mathematical model. Completing the collected data, the association between the possibility of business process performance and an operation of particular appliances can be used to present a measurement method to assess the computer system reliability in the context of business processes.

References

1. 25777:2008 – Information and communication technology continuity management (2008)
2. British standard BS 25999-1: Business Continuity Management. Code of Practice (2006)
3. British standard BS 25999-2: Specification for Business Continuity Management (2007)
4. PN-ISO/IEC 27001:2007 - Zarzadzania bezpieczenstwem informacji (2007)
5. Tannenbaum, A.S.: Computer Networks, 4th edn. (2003) ISBN: 0-13-066102-3
6. Grzywak, A., Domanski, A., Domanska, J., Rostanski, M.: Sieci Komputerowe. WKiL (2007) ISBN: 978-83-88936-34-0
7. Blokdijk, G., Brewster, J., Menken, I.: Disaster Recovery and Business Continuity IT Planing, Implementation, Managing and Testing of Solutions and Services Workbook. The Art of Service (2008)
8. Myers, K.N.: Business Continuity Strategies. Protecting Against Unplaned Disasters. John Willey & Sons, Inc., New Jersey (2006)
9. Anderson, R.: Inzynieria zabezpieczen: Wydawnictwa Naukowo-Techniczne, Warszawa (2005)
10. Gibb, F., Buchanan, S.: A Framework for business continuity management. International Journal of Information Management 26, 128–141 (2006)
11. Zawila-Niedzwiecki, J.: Metoda TSM-BCP projektowania rozwiźzan zapewniania ciaglosci dzialania organizacji. Instytut Organizacji Systemow Produkcyjnych Politechniki Warszawskiej (2007)
12. U.S Departament Of Defence: Trusted Computer System Evaluation Criteria, DOD 5200.28 (1985)
13. Wilson, E.: Microsoft Windows Scripting with WMI. Microsoft Press (2006)

Chapter 14
Computer Support in Business Continuity and Information Security Management

Andrzej Bialas

Abstract. The chapter features the concept of joint implementation of two widely used standards, BS 25999 concerning business continuity and ISO/IEC 27001 – information security, within one integrated management system. Business continuity is understood as a strategic ability of an organization to react to incidents and disturbances in its business functioning and to mitigate losses in case these harmful factors occur. Information security, in turn, is related to the protection of information integrity, availability and confidentiality. The chapter presents the assumptions and recently achieved results of a specific targeted project whose objective is to develop a computer-supported management system for organizations which set a lot of store by the continuity of business processes and information security. The works on the system model were summarized, including a feasibility study concerning different aspects of software which is developed on the basis of these models.

14.1 Introduction

The chapter features the issues concerning the implementation of two basic groups of security standards:

1. BS 25999 [1], [2] on business continuity (BCMS – Business Continuity Management Systems), and
2. ISO/IEC 27001 [3], [4] on information security (ISMS – Information Security Management Systems),

in the form of one, integrated software called OSCAD. The basic objective of the software is to support management in these two aspects of the organization's functioning.

Andrzej Bialas
Instytut Technik Innowacyjnych EMAG
ul. Leopolda 31, 40-189, Katowice
e-mail: a.bialas@emag.pl

A. Kapczynski et al. (Eds.): Internet - Technical Develop. & Appli. 2, AISC 118, pp. 155–169.
springerlink.com

Business continuity is understood as a strategic ability of the organization to:

1. plan its reactions and react to incidents and disturbances in its business functioning so that its operations could be continued on an acceptable, previously determined level,
2. mitigate losses in case such harmful factors occur.

Business continuity is related to business processes carried out in an organization. Information technologies can support the execution of these processes to a certain extent. Information security is related to the protection of the integrity, availability and confidentiality of information which is generated, processed, stored, and transferred when these business processes are underway. Both issues are of similar character and penetrate into each other's realms in organizations. For that reason it is justifiable to integrate these two management systems, though, for the OSCAD project as such, it was assumed that business continuity should be given priority over information security.

Business continuity is often based on proper functioning of IT infrastructure which supports the execution of business processes. Still, apart from IT-related events, there are many other factors which can disturb business continuity, such as: technical damages; random disruption events; catastrophes; disturbances in the provision of services, materials and information from the outside; organizational errors; human errors, deliberate human actions, etc.

The continuity of business processes and proper information protection in an organization is the basis to successfully compete on the market. To secure business continuity is extremely important and, at the same time, difficult for organizations:

1. which are part of complex supply chains, have expanded co-operation links,
2. which work according to the Just-in-time strategy where the volume of stock needed for production is strictly optimized,
3. whose business processes depend strongly on efficient operations of IT systems,
4. which have on-line services systems that are one of the most advanced forms of IT applications supporting business processes.

The implementation of a business continuity management system in an organization results in the development of a suitable business structure that will have higher resistance to threats and will enable to react to incidents in such a way that negative impact for business continuity will be of the smallest possible degree. The recommendations and requirements concerning business continuity are included in standards [1], [2].

The chapter features the concept of the OSCAD system developed at the Institute of Innovative Technologies EMAG within a project co-financed by the Ministry of Science and Higher Education.

Currently, the project is at the phase of assessing the efficiency and feasibility of management processes models, which allows to move to the phase of integration and implementation.

14.2 Basic Standards of the OSCAD Project

The OSCAD project is based on the BS 25999 standard. Its first part [1] contains definitions related to business continuity along with a set of best practices and recommendations in the range of business continuity management. The second part is the specification of requirements which are the basis to develop and certify business continuity management systems.

It is important to note that business continuity management systems, just as similar systems (information security management-, quality management-, IT services-, or health and occupational safety systems), are developed based on the Deming circle (cycle) [5], also known as the PDCA cycle (Plan-Do-Check-Act), as it groups management processes of a system into four steps:

1. Plan – processes related to working out the plan of a project concerning the management on the basis of a previously elaborated method;
2. Do – processes concerning a trial implementation of the plan and implementation of the method;
3. Check – evaluation processes which check whether the new method gives better results;
4. Act – if the new method brings better results it should be considered as a valid procedure, standardized, its functioning should be monitored and continuously improved.

The Act phase detects incompatibilities of the management system with the current situation of the organization and its environment. For example, changes in the method of business processes execution or new forms of threats make it necessary to adapt the management system to these new conditions. Changes in the system have to be planned, done, checked and implemented for exploitation. This way the PDCA system executes its ability to self-adaptation.

The Deming circle ensures controlled implementation of a management system, its unassisted adaptation to new challenges as well as improvement based on measurable indicators.

For the OSCAD project a three-layer architecture was adopted [6]:

1. organizational layer – includes management processes, system planning and maintenance, development of organizational structure, elaboration of procedures and operational plans, training and awareness raising processes, internal and external communication, preparation and analysis of statistical data; the organizational layer is supported by software;
2. logical layer – is executed by software and encompasses computer-based decision support, Business Impact Analyses (BIA), risk management, data base maintenance, task management, records of system operations, monitoring incidents and threats and automatic generation of warnings, collecting procedure patterns, reporting;
3. technical layer – includes the execution of communication channels, co-operation with computer systems working in a network, development and maintenance of threat reporting- and warning points; the physical layer is part of the developed

software and is supported by its environment, including IT infrastructure, business applications and physical infrastructure.

The business continuity management system [1], [2] was integrated with an information security management system according to [3], [4] based on the concept of the so called Integrated Security Platform (ISP) presented in [7]. ISP makes use of a common Configuration Management Database (CMDB) applied in IT services management systems [8], [9], [10]. The ISP platform ensures the management systems integration from the IT point of view, while BS PAS 99 [11] provides recommendations for the organizational and procedural aspect. They assume separation of a common part of the organization's management systems and its uniform, integrated implementation. Specific elements are implemented for each system separately. The management systems are integrated in order to lower the costs and improve management efficiency.

14.3 The Objectives of the OSCAD Project

The basic objective of the OSCAD project is to develop an integrated management system for business continuity and information security. The system is to ensure:

1. fluent, monitored functioning of an organization in crisis situations when business continuity is disturbed or information security breached,
2. ability to mitigate the impact of business continuity disturbances or information security breaches,
3. ability to restore business processes to their original form after different types of incidents.

It comes down to the elaboration of certain methods and tools, including knowledge how to use these methods and tools. Within the OSCAD project the following will be elaborated:

1. methodology to build an open, frame-type, integrated management system for business continuity and information security in the form of standard modules (patterns),
2. methodology to implement the system based on the identification of the needs and requirements of the organization which plans to implement the system and on the adaptation of standard modules (patterns) to the form of final modules on the basis of these needs and requirements,
3. software supporting the process of implementation and, later, exploitation of the integrated management system for business continuity and information security,
4. software to build a system that will collect, analyze and provide statistical information about threats, vulnerabilities and incidents disturbing the organization's business processes.

The execution of the project refers to previously conducted research on information security [12], [13], [14], [15], [16], [17], [18], risk management [19], [20] and security ontology [21].

14.4 The Range of Computer Support of Operations Related to the Management of Business Continuity and Information Security

Within the project the needs of future users of the software were analyzed, the requirements resulting from standards were determined, and the trends in the development of methods and technologies were reviewed. On this basis the assumptions for the project and models were elaborated. The division was made between the functions performed automatically by the software and those executed by people involved in the management of business continuity and information security. A feasibility study was performed for key modules and software elements.

An extremely important issue was to determine the range of computer support of business continuity and information security management processes. Here, similarly to Computer Aided Design/Manufacturing/Engineering (CAD/M/E), special attention was paid to operations that are complicated, repeatable and laborious. Their automation allows to get information for decision making more quickly and to use previously elaborated documents and collected data in new operations during the exploitation of business continuity and information security management systems (BCMS/ISMS). Moreover, the automation of management processes increases their quality and efficiency. After analyzing BCMS/ISMS processes described in standards [1], [2], [3], [4] there were a series of activities selected for automation, of which the most important are the following:

1. generating different documents of a BCMS/ISMS management system based on templates; managing documents, including their filtering and presentation;
2. specification and analysis of business processes,
3. task management within BCMS/ISMS management processes,
4. communication,
5. assets inventory of the organization,
6. BIA analysis, risk analysis,
7. audits and reviews,
8. planning and timetabling,
9. incident management,
10. business continuity plans and their verification,
11. managing the personnel that take part in management processes,
12. correction and improvement actions, change management,
13. records management,
14. training and awareness raising actions.

14.5 General Concept of the OSCAD Software

The OSCAD software system is a complex system. There are two major components in it:

1. the basic OSCAD component which manages business continuity and information security and communicates with similar systems and other components,
2. the OSCAD-STAT component responsible for collection, analysis and provision of statistical information about incidents; this component also plays a role of a thematic portal.

Figure 13.1 shows the basic OSCAD component (system) against external elements that take part in the business continuity and information security management process. The following were distinguished [22]:

1. internal modules group of the OSCAD system; the modules are responsible for the execution of basic tasks in the range of management support; the events and incidents management modules were distinguished from the group with a view to show their specific relations; Figure 13.2 presents the internal modules of the OSCAD software in a more detailed way,
2. events and incidents management modules as central elements of the system, collecting information about events and allowing their assessment and proper classification,
3. user interface in the form of a group of panels co-operating with main users of the system,
4. notification subsystem which is also responsible for automatic collection of information about events generated by different types of devices working in the software environment,
5. data base system which stores all information about the organization, its security and business continuity, about roles played in the organization, events, incidents and undertaken actions, as well as about management processes as such.

The events management module and incident management module enable to exchange information with external systems which collect information about events, such as: organization's assets management systems, systems for monitoring IT and physical infrastructure, facility protection systems, including fire protection systems, etc. Direct notification about events was envisaged too by means of e-forms, along with communication with other OSCAD systems working in other organizations or in the organization's departments. After the event is registered, its character is assessed. This way some events become incidents. The user interface module was distinguished which allows the users – actors performing different management functions to communicate with the OSCAD system (they report events, assess them, react or conduct different analyses).

The notification system consists of two basic elements: one that collects information and the other that distributes it. The first one gets information from its own notification system (burglar, fire, and safety alarm systems optionally equipped with alarm buttons "Damage" or "Accident" on the production line). It is possible to collect data automatically from different types of sensors or external computer systems, e.g. Enterprise Resource Planning (ERP) systems, IT infrastructure or technical infrastructure monitoring systems. The module that distributes information has to inform (by means of all available channels – SMS, e-mail, telephone, internal radio communication system) the people who have to be informed about the event according to proper operation procedures.

Information about the system configuration, roles, dictionaries, incidents, business processes, risk analysis results, audit results, undertaken actions, measures and indicators, etc. are stored in the OSCAD system data base. This data base is the central part of the system that plays an integrating and supporting role for all internal modules.

Due to security reasons, it is assumed to run a stand-by OSCAD system and to replicate the data base in order to use them both in situations when the main system is not available. The OSCAD software is also able to exchange information with similar systems within a supply chain. This concerns particularly the possibility of mutual warning about incidents.

OSCAD-STAT is a central statistical system that exchanges information with the OSCAD systems. It will be described further in the chapter (Fig. 13.3).

14.6 Review of Modules Responsible for Business Continuity and Information Security Management

The diagram in Figure 13.2 (UML component diagram) presents internal modules of the OSCAD system and their mutual relations (as interfaces in UML). To make the diagram easy to read, only the most important relations with modules were shown. The system management module supports the execution of tasks by other modules and concerns proper configuration of the OSCAD system, preparation of pre-defined data, including dictionaries, basic information about the organization, risk analysis parameters (business loss matrix, acceptable risk level), and the roles of people who take part in business continuity and information security management processes. It is assumed that pre-defined implementation profiles can be uploaded.

The operation of the business continuity management system is based on the identification of business processes and services which are critical for the organization. Then, certain actions are undertaken to minimize the risk of an incident which threatens the business continuity and to work out methods which, after the incident occurs, enable to restore access to the organization's critical processes and services as quickly as possible.

Therefore, before a BCMS system is planned, it is first necessary to get familiar with the way the organization functions, with its business processes and mutual internal dependencies (within particular business processes and between processes) and external dependencies (relations with suppliers and contractors, environmental conditions). The OSCAD system enables to store information about the organization and its business processes which are then used in successive management processes (e.g. risk analysis). All actions undertaken to provide business continuity in the organization are focused around business processes.

The assets inventory module is responsible for storing information about assets and asset groups related both to the business continuity- and information security management system. Three subsets of data related to the given asset were distinguished:

1. common data – used in each system (BCMS/ISMS) and comprising, among others: asset name, asset kind, asset owner, asset assignment to certain processes,
2. data required by the BCMS system, i.e. related to business continuity, such as: determining the asset criticality with respect to its availability, time and expenditure needed to restore the asset,
3. data required by the ISMS system, i.e. related to information security, such as: the asset assignment to a certain information group and this way determining the importance of the asset.

According to the assumptions, the module will enable to collect, modify and delete information about assets placed in the data base.

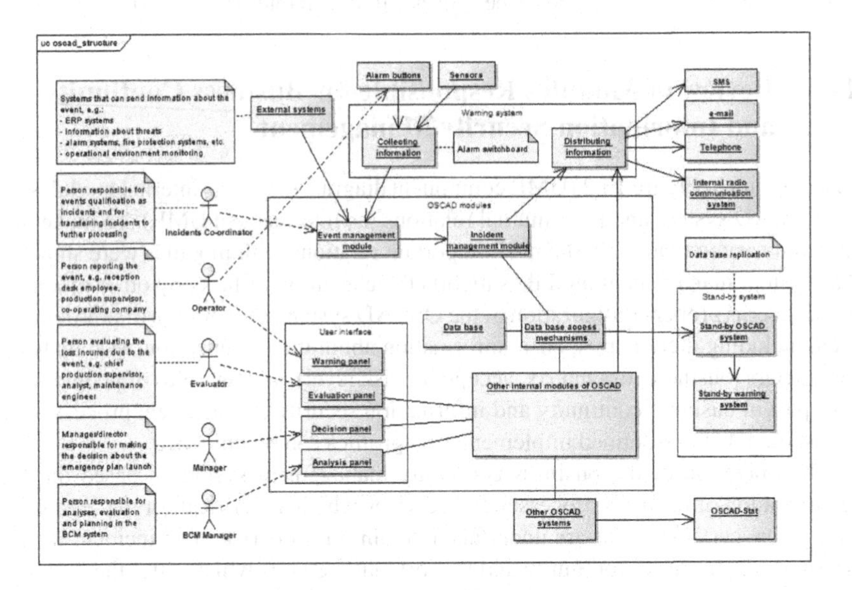

Fig. 14.1 Diagram of the OSCAD system functioning

14.7 Computer Support in Business Continuity and Information Security Management

The risk analyzer module is one of the key modules of the OSCAD system. It allows to direct the implementation of the system in the organization's operational environment. The module supports the execution of activities related to risk management, such as:

1. configuration of risk analysis parameters – determining the acceptable risk level, determining the business loss matrix, the method to calculate the importance of the process,

2. Business Impact Analysis (BIA) of the loss of such parameters as availability, integrity of a process, service or information as well as information confidentiality (important in the case of an information security management system),
3. determining criticality of processes based on the BIA analysis results,
4. determining parameter value of Maximum Tolerable Period of Disruption (MTPD) and Recovery Time Objective (RTO),
5. collecting information about threats and vulnerabilities for processes, information groups (possibility to select threats and vulnerabilities defined in bases/dictionaries or to enter new ones),
6. risk assessment with respect to existing security measures,
7. analysis results reporting,
8. preparing a risk treatment plan: risk acceptance or selection of security measures with respect to implementation costs (the so called economic calculation).

The BIA analysis is required by different management systems, also by a business continuity management system described in BS 25999. Additionally, BIA can be part of the risk analysis process of information security management systems described by the ISO/IEC 27001 standard.

The events and incidents management module, whose connections are presented in Figure 13.1, is responsible for the execution of the following tasks in the BCMS system:

1. event registration, incident classification and registration,
2. preliminary evaluation and selection of proper proceedings,
3. initiation of rescue operations to protect lives and health of employees and clients,
4. problem analysis and start-up of a proper Business Continuity Plan (BCP),
5. communication with interested parties and co-operation with contractors,
6. providing accessibility of the basic and stand-by locations of incident management,
7. providing all required documents,
8. closing the incident,
9. reporting,
10. lessons learnt from incidents.

An important task within the D phase of the Deming circle is the elaboration and maintenance of business continuity plans (BCP) which determine:

1. how the organizations aims to ensure business continuity,
2. how business processes will be restored to their normal condition.

The BCP elaboration and maintenance module supports three basic activity groups: elaboration, start-up and testing of the BCP plan. It was assumed that plans are created for processes which, as a result of the conducted risk analysis, have the "critical" attribute assigned. The BCP plan points at:

1. assets necessary to start-up and perform the plan,
2. plan execution environment – basic and stand-by locations in the organization,

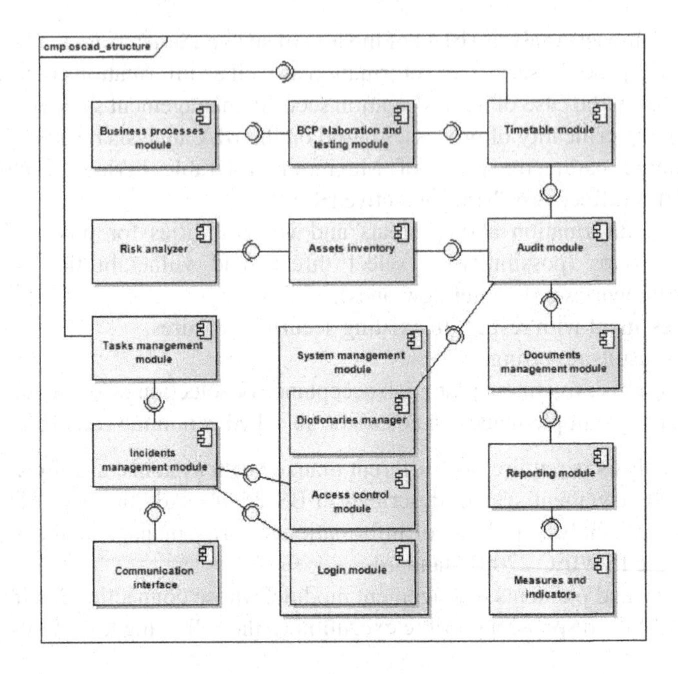

Fig. 14.2 Basic components of the OSCAD system

3. contact list of people involved in the plan execution,
4. operations to be carried out.

Tests can be of different character, depending on the kinds of operations undertaken within the plan. Tests planning is supported by the timetable module.

The task management module has a universal character. It enables to define tasks to be performed by particular users of the systems and to control the performance. The tasks can be generated directly from this module or created from the level of other modules as a result of operations carried out in the module. The sample operations can be the following:

1. operations implied by the risk analysis,
2. incident-handling operations registered in the incident management module,
3. operations resulting from actions planned in the timetable module,
4. operations resulting from the elaborated BCP plans, in case the a particular plan has to be launched.

The basic functions of the module are planning, ordering, supervising the performance and reviewing tasks. Apart from the management of business continuity and information security in the organization, the tasks can refer to any other aspect of the organization's operations. The only condition is that the user to whom the task performance is assigned should have an account in the OSCAD system. The module contains functions notifying the users about events related to the tasks flow in the system (ordered, performed, overdue) and allows to plan the tasks. The module functionality is the basis for the functionality of the incidents management module and supplements the functionalities of many other modules of the system.

The audit module supports audit management in the OSCAD system, i.e. it collects information about the execution of each audit, supports audit reports preparation, helps in audit approval by people who supervise it. Audit planning as such is handled in the timetable module.

The timetable module supports to plan management operations undertaken at a certain time horizon within different modules.

Information about the BCMS/ISMS system collected during its implementation and kept during exploitation can be presented as text, graphics or exported with the use of popular formats. This job is done by the reports module. The following reports are expected as minimum:

1. information about the organization and its business processes,
2. information about the organization's assets involved in the execution of the OSCAD system,
3. summary of the risk analysis process,
4. summary of planned tasks which are part of timetables,
5. information about BCP plans,
6. summary of audit results,
7. summary of preventive and corrective actions,
8. values of measures and indicators used to continual improvement of the BCMS/ISMS system,
9. summary of different tasks execution,
10. statistics related to incidents.

The documents management module is responsible for registration, version control, circulation, confirmation, search, etc. of BCMS and ISMS systems documents according to the requirements of standards. Documents and their templates are attached to the system and stored in the data repository. Sample documents are: "Business continuity management system policy", "Description of risk assessment methodology and risk acceptance criteria", or "Records supervision procedure".

The communication module is responsible for information exchange between external systems and appointed persons. The communication interface automatically performs the phases of event detection and notification for the incident management module. Events registration can be done "manually" (entered by a person who gives information about the event) or automatically by means of an XML template generated and sent by the system or by external devices, including other OSCAD-type systems.

The external systems that are currently taken into account can be divided with respect to the functions they perform:

1. Enterprise Resource Planning (ERP) systems,
2. IT infrastructure, including IT services monitoring systems,
3. technical infrastructure monitoring systems,

4. building automation systems,
5. burglary or fire alarm systems.

The communication interface will provide connection with external systems through most recognized communication protocols.

The measures and indicators module stores the efficiency parameters of the management system. This allows to conduct periodical analysis and make decision about actions to improve the system.

The dictionaries module results from the need to apply different types of predefined lists used by the software. Dictionaries allow to easily adapt the OSCAD system to the organization's profile. They also contain specific elements, such as requirements that are subject to audits. These can be the requirements of standards, laws or internal regulations of the organization. The dictionaries also encompass threats, vulnerabilities and security measures specific for the given organization.

14.8 Review of Modules Responsible for Statistical Information Management and Other Auxiliary Modules

Figure 13.3 features basic elements of the OSCAD-STAT system.

It was assumed that OSCAD-STAT should consist of two independent functional modules. The first one, OSCAD-STAT ANALYSIS is responsible for:

1. data exchange with OSCAD systems working in different organizations,
2. preparation of statistical data,
3. giving access to corrections and new versions of the OSCAD system (option).

The OSCAD-STAT PORTAL module is responsible for:

1. presentation of statistical data collected by the OSCAD system and other supporting systems,

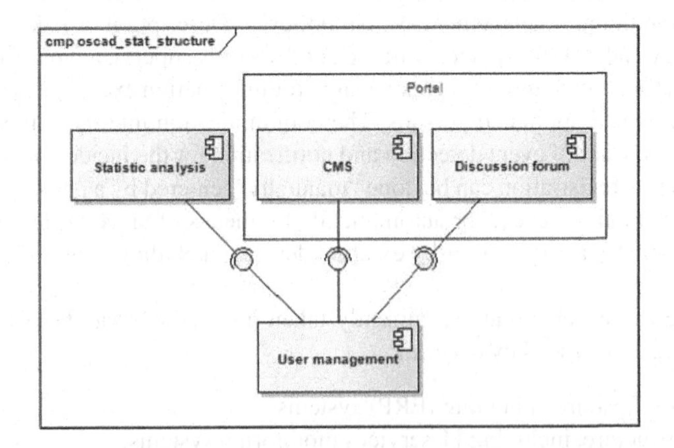

Fig. 14.3 Basic elements (UML components) of the OSCAD-STAT system

2. dissemination of knowledge concerning business continuity and knowledge accumulated within the conducted project,
3. giving access to a platform where there is exchange of experiences on information security and business continuity between all interested parties.

Additionally, there are works conducted on the OSCAD-REDUNDANCY module which will be responsible for providing access to a complex information and communication platform. The platform will enable to start up the redundant OSCAD system which will be run only in crisis situations when the basic system is not accessible by the organization.

14.9 Conclusions

The chapter concerns assumptions and first results of the specific targeted Project "OSCAD – open, frame-type, integrated management system for business continuity and information security" which has been carried out for a year at the Institute of Innovative Technologies EMAG.

The project concerns technical and organizational issues related to the construction of business continuity management systems in organizations. The OSCAD project has the following innovative elements:

1. open character of the system achieved due to withdrawal from a dedicated solution and the development of a set of standard modules (patterns) along with the methodology of their adaptation and implementation according to the organization's needs;
2. development of an advanced tool supporting the implementation and maintenance processes of a BCMS system;
3. integration with an information security management system ISMS (ISO/IEC 27001);
4. possibility to conduct a statistical analysis of events and reliability analysis of IT systems functioning;
5. possibility to exchange information between different organizations;
6. easy integration with other management systems co-existing in the organization;
7. support for organizations working within supply chains.

The system configurability and scalability are achieved thanks to standard and parameterized modules. The system can be constructed by means of selecting particular modules from the set of pre-defined patterns and then customizing them to the needs of the organization. These needs are identified in the first phase of the customization and implementation methodology.

Business continuity issues are of key importance for modern businesses and public administration. However, the solutions have to be implemented and exploited with respect to the needs and costs analysis. Thanks to the applied risk management method, it is possible to control the relations between the achieved business continuity indicators and the costs to achieve them. This has a positive impact on the efficiency of the business continuity management process in the organization.

The project results are chiefly dedicated to:

1. organizations which are elements of the critical infrastructure of the country (power engineering, production and distribution of fuels, telecommunications, etc.),
2. financial institutions (e.g. in the sectors of insurance or banking),
3. organizations offering e-services,
4. public administration (government- or local government level),
5. organizations representing the sectors of health services and higher education
6. organizations involved in the protection of groups of people,
7. other commercial and industrial companies.

The running project is directed mainly to different forms of business and the needs of the administration. Additionally, after certain modifications, it can be used, for example, in systems which support crisis management.

In the course of the prototypes development special focus will be on versions dedicated to specific applications. This requires close co-operation with specialists of particular domains.

The OSCAD system can be used in any organization which plans to implement a BCMS or ISMS system compliant with standards. This implementation does not have to be completed with a formal certification process, however, the use of the OSCAD system will facilitate the organization's preparation to such certification.

It is important to note that thanks to reusability the maintenance of the system is easier because the organization can use the data collected during implementation many times, adapting them to the occurring changes. The challenge for the OSCAD team was to achieve advantages similar to those derived from computer systems supporting development or production.

References

1. BS 25999-1:2006 Business Continuity Management – Code of Practice (2006)
2. BS 25999-2:2007 Business Continuity Management – Specification for Business Continuity Management (2007)
3. PN-ISO/IEC 27001 – Technika informatyczna – Techniki bezpieczenstwa – Systemy zarzadzania bezpieczenstwem informacji – Wymagania
4. PN-ISO/IEC 17799:2007 – Technika informatyczna – Techniki bezpieczenstwa – Praktyczne zasady zarzadzania bezpieczenstwem informacji
5. `http://pl.wikipedia.org/wiki/Cykl_Deminga`
6. Bialas, A.: Komputerowo wspomagany system zarzadzania ciagloscia dzialania – zalozenia projektu. In: Materialy konferencyjne EMTECH 2010 – Zasilanie, Informatyka Techniczna i Automatyka w Przemysle Wydobywczym – Innowacyjnosc i Bezpieczenst wo, Ustron, May 19-21, pp. 29–37 (2010)
7. Białas, A.: Development of an Integrated, Risk-Based Platform for Information and E-Services Security. In: Górski, J. (ed.) SAFECOMP 2006. LNCS, vol. 4166, pp. 316–329. Springer, Heidelberg (2006)
8. IT Infrastructure Library, `http://www.itil.co.uk`

9. ISO/IEC 20000-1:2005 Information technology – Service management – Part 1: Specification
10. ISO/IEC 20000-2:2005 Information technology – Service management – Part 2: Code of practice
11. BS PAS 99:2006 Specification of common management system requirements as a framework for integration
12. Bialas, A.: Bezpieczeñstwo informacji i usşug w nowoczesnej instytucji i firmie. Wydawnictwa Naukowo-Techniczne, Warszawa (2006, 2007)
13. Bialas, A.: Using ISMS concept for critical information infrastructure protection. In: Balducelli, A., Bologna, S. (eds.) Proceedings of the International Workshop on "Complex Network and Infrastructure Protection – CNIP 2006", ENEA – Italian National Agency for New Technologies, Energy and the Environment, Rome, March 28-29, pp. 415–426 (2006)
14. Bialas, A.: The ISMS Business Environment Elaboration Using a UML Approach. In: Zieliñski, K., Szmuc, T. (eds.) Software Engineering: Evolution and Emerging Technologies, pp. 99–110. IOS Press, Amsterdam (2005) ISBN: 1 58603-559-2
15. Bialas, A.: A UML approach in the ISMS implementation. In: Dowland, P., Furnell, S., Thuraisingham, B., Wang, X.S. (eds.) Security Management, Integrity, and Internal Control in Information Systems, IFIP TC-11 WG 11.1 \& WG 11.5 Joint Working Conference, pp. 285–297. Springer Science + Business Media, New York (2005) ISBN-10:0-387-29826-6
16. Bialas, A.: Development of the Information Security Management System for the Polish Mining Sector. In: Mechanizacja i Automatyzacja Gornictwa, Katowice. Centrum Elektryfikacji i Automatyzacji Gornictwa EMAG, vol. (8), pp. 34–41 (2007)
17. Wartak, A., Lisek, K.: Walidacja systemu zarzadzania bezpieczenstwem informacji dla sektora wegla kamiennego. In: Mechanizacja i Automatyzacja Gornictwa, Centrum Elektryfikacji i Automatyzacji Gornictwa EMAG, Katowice, vol. (10), pp. 39–46 (2007)
18. Styczen, I., Baginski, J.: Oprogramowanie wspomagajace system zarzadzania bezpieczenstwem informacji. In: Mechanizacja i Automatyzacja Gornictwa, Centrum Elektryfikacji i Automatyzacji Gornictwa EMAG, Katowice, vol. (11), pp. 22–29 (2007)
19. Bialas, A., Lisek, K.: Integrated, business-oriented, two-stage risk analysis. Journal of Information Assurance and Security 2(3) (September 2007), `http://www.dynamicpublishers.com/JIAS`
20. Bialas, A.: Security Trade-off – Ontological Approach. In: Akbar Hussain, D.M. (ed.) Advances in Computer Science and IT, In-Tech, Vienna-Austria, Vukovar-Croatia, pp. 39–64 (2009) ISBN 978-953-7619-51-0
21. Bialas, A.: Ontological Approach to the Business Continuity Management System Development. In: Arabnia, H., Daimi, K., Grimaila, M.R., Markowsky, G. (eds.) Proceedings of the 2010 International Conference on Security and Management (The World Congress In Applied Computing – SAM 2010: July 12-15, Cityplace Las Vegas, country-regionUSA), vol. II, pp. 386–392. CSREA Press (2010) ISBN: 1-60132-159-7, 1-60132-162-7 (1-60132-163-5)
22. Zbiorowa, P.: Raporty projektu celowego pn. Komputerowo Wspomagany System Zarzadzania Ciagloscia Dzialania – OSCAD, Instytut EMAG (2010-2011)

Chapter 15
High Availability Methods for Routing in SOHO Networks

Maciej Rostanski

Abstract. Today, even for small businesses, data and IT services must be cloud-ready and available at any time, from any location, by anyone. End-user appreciate (and increasingly, expect) of always-on business resources and non-disruptive operations. Improving data backup and recovery, and business continuity / disaster recovery are among top 10 IT priorities. As gateway is often the most vulnerable network element, this article focuses on an Internet protocol called VRRP (Virtual Router Redundancy Protocol) utilization as it enables the ability to have backup routers when using a statically configured router at some point of a local area network.

A redundant, cost-effective network block instead of one single router is presented, along with test topology and some guidelines. Test results are included and discussed.

15.1 Introduction

In reality world, typical business computer network is not funded enough to be equipped with complex solutions. The goal is simple - the network has got to be deployed and maintained at minimum cost. Most of time, typical SOHO network would consist of one edge router, single physical server and minimized second-layer network infrastructure. Moreover, IT departments are under constant pressure of cost cutting - or cost avoidance, if the head of IT is not too much concentrated on the infrastructure lower level. The difference is that cost avoidance requires some initial strategic capital outlay, which is recovered through reductions in operational costs, productivity, or other savings realized through deploying the technology.

Maciej Rostanski
Academy of Business in Dabrowa Gornicza, Department of Computer Science
ul. Cieplaka 1c, 41-300 Dabrowa Gornicza, Poland
e-mail: mrostanski@wsb.edu.pl
http://www.wsb.edu.pl

A. Kapczynski et al. (Eds.): Internet - Technical Develop. & Appli. 2, AISC 118, pp. 171–178.
springerlink.com

Quoting [5]:

> Approximately 70 percent of IT investment is spent on "keeping the lights on" - providing basic connectivity and security - leaving precious little for strategic projects to enhance the business. While such tactical thinking is understandable, it also contains many pitfalls.

On the other hand, modern network technologies must support today's environments. They must help deliver IT as a service that is consumed when needed; data and IT services must be cloud-ready and available at any time, from any location, by anyone. Virtualization of server processing power, data storage capacity, and network bandwidth are becoming ubiquitous. The result is end-user appreciation (and increasingly, expectation) of always-on business resources and non-disruptive operations. According to [1], improving data backup and recovery, and business continuity / disaster recovery were among top 10 IT priorities for next 18 months, in 2010.

One of the key issues of network reliability even for small businesses, is extranet availability. For SOHO segment, this would most of the time mean Internet access availability (and performance), as it is used for data acquisition, or for example VPNs and remote working. It is obvious that the applications provided by the company intranet or the Internet impact employee productivity, customer experience, and partner interactions with the company. Most of the time, access to services outside local area network is crucial.

In order to minimize downtime, one of the strategies for the network reliability is introducing redundancy to the key elements on many levels. For example, links in physical layer (multiple network adapters and connections), switching on link-layer level (multiple switches, multi-connected with STP protocol family or its modern successor, TRILL protocol) or IP routing (using multiple routers).

The problem for small businesses is lack of sufficient IT resources for maintaining and administering complex network solutions, not to mention limited budget, as described before. Author presents simple and effective routing redundancy topology that doesn't affect configurations of other network elements and clients.

Although there are quite a couple such solutions at network layer, such as HSRP, GLBP or CARP, this article focuses on an Internet protocol called VRRP (Virtual Router Redundancy Protocol), as it is both: (1) common standard in low-budget solutions, (2) is easily configured, and (3) is scalable as it enables the ability to have more backup routers when using a statically configured router at some point of a local area network. This ability is an outcome of specifying, as stated in [4]:

> an election protocol that dynamically assigns responsibility for a virtual router to one of the VRRP routers on a LAN.

15.2 Routing Redundancy with VRRP

The key to reliable and fault-tolerant infrastructure development is to define weak elements (elements, that should not fail, or fail too often), classify them with priorities, business goals and overall importance in mind, and apply different techniques of high availability, among them:

- Clustering;
- Load balancing;
- Redundancy protocols;
- Failover;
- Configuration synchronization;

15.2.1 Redundancy Solutions for Routing

Redundancy techniques differ depending on the element to be more fault-tolerant. On abstract level, though, those could be described as:

1. *Passive* way of introducing redundancy, achieved by having an element ready for replacement. For example, an IT department may have some spare switches or routers, ready to be swapped with failed equipment.
2. *Active* redundancy, which means protected equipment and its backup element are working simultaneously. Redundant power units, two domain controllers or RAID 5 disk array are examples of an active redundancy.

As nowadays networks utilize practically IP protocol and routers and switches (layer-two and layer three) as the only active elements, bringing fault-tolerance to routed and switched IP networks is the main point of interest, especially in SOHO segment. In physical and link layers, Ether-Channel and STP protocols (PVST, MST) are most common techniques. As for network layer (fault-tolerant routing), most often there is no fault-tolerance strategy.

As an outcome, gateway is the most vulnerable network element in these situations. That is the reason, as mentioned earlier in 1.1, routing redundancy protocols were developed, such as VRRP, to provide the ability to create backup routers, ready to replace master gateway during failure.

15.2.2 VRRP Protocol

VRRP is designed as a *passive* redundancy method - one of the routers takes care of the traffic, backup routers are not used in most common topology, however using carefully designed logical and physical topology, one can achieve load balancing with VRRP routers and create *active* redundant solution. Standardized by IETF, VRRP is described thoroughly in [3]. Current version, VRRPv3 supports IPv6 as a network protocol.

Availability increase is achieved by suggesting "virtual router" (representation of master router and backup routers, acting as a group) as a default gateway for hosts on a subnet, instead of real router address. If there is a failure, the next usable router is elected master router.

For its work, VRRP leverages the ARP protocol mechanisms, populating client ARP tables with virtual router address entries - whenever a client sends an ARP request about virtual IP address, master router responds.

Typical router configuration is presented on 15.1. In this topology, there is a backup route with another ISP to the external network. In case of R1 router failure, R2 becomes master (active) router.

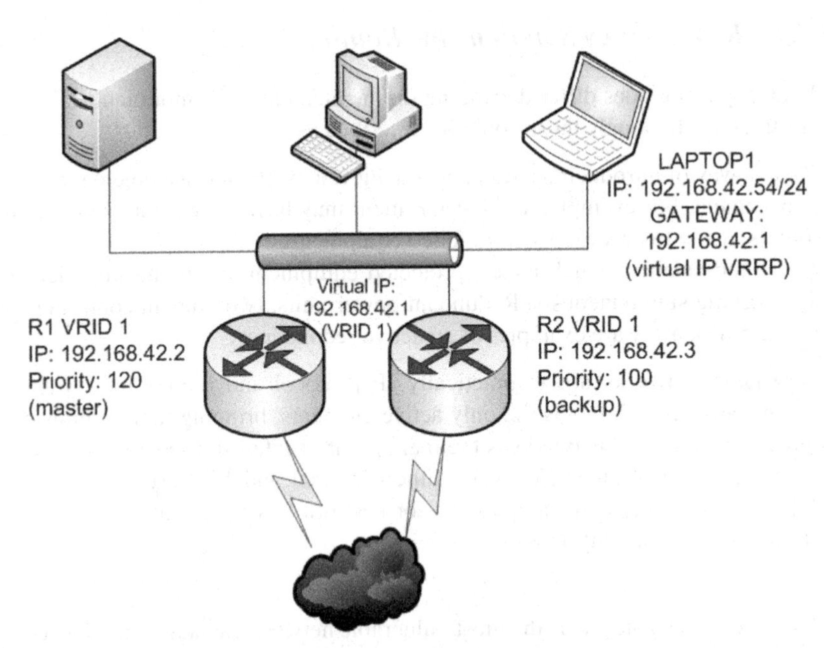

Fig. 15.1 Basic VRRP configuration. R1 is the master router, R2 is the backup router. This is passive redundancy - R2 is not used until R1 suffers a failure.

If the link fails, R1 needs to become backup router - this is accomplished using either interface tracking or VRRP sync group.

Interface Tracking

You can configure VRRP to track an interface that can alter the priority level of a virtual router for a VRRP virtual router. When the IP protocol state of an interface goes down or the interface has been removed from the router, the priority of the backup virtual router is decremented by the value specified in the priority-decrement argument. When the IP protocol state on the interface returns to the up state, the priority is restored. The problem with this solution is VRRP itself doesn't provide this mechanism, it's got to be vendor supported.

VRRP sync Group

Another solution is VRRP sync group. Interfaces in a sync group are synchronized such that, if one of the interfaces in the group fails over to backup, all interfaces in the group fail over to backup. For example, on 15.1, if one interface on a R1 fails, the whole router should fail over to a backup router. By assigning all the interfaces on the master to a sync group, the failure of one interface will trigger a failover of all the interfaces in the sync group to the backup configured for the interface.

15.3 Simple Solution for Redundancy

The idea is to create a redundant, cost-effective network block instead of one single router. For flexibility, configuration and cost-effectiveness, we used software router Vyatta (community edition)[7]. Vyatta was operating on two identical basic server units R1 and R2, connected with switches on both sides.

For research purposes, we created a scripted environment, simulating subsequent failures of R1.

15.3.1 Topology

The topology is presented on 15.2. VRRP is set on both sides of routers R1 and R2, so for the rest of the network they appear as one router (with IP addresses 192.168.1.254 and 192.168.42.1). No NAT mechanisms are in use, although possible; the idea is to provide the ability to communicate between clients on both sides of the intranet. With VRIDs and virtual addresses on both sides, there is no need of

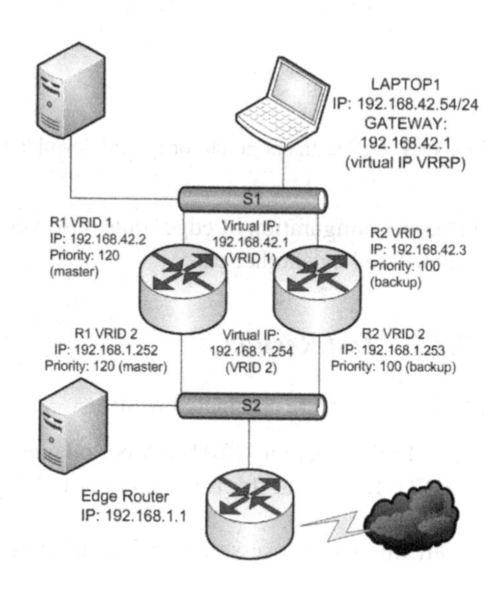

Fig. 15.2 Tested simple topology for IP routing redundancy with VRRP. There is a RIP protocol enabled for routers to exchange information about active routes, STP protocol as default on switches. Router R1 is the master router and is being shut down for testing purposes, respectively.

Table 15.1 Connectivity test results

Action	Min break [s]	Max break [s]	Mean [s]
R1 off-line (fail)	4.35	17.05	6.43
R1 on-line (repair)	28.7	39	34.1

(Results of 50 ICMP tests)

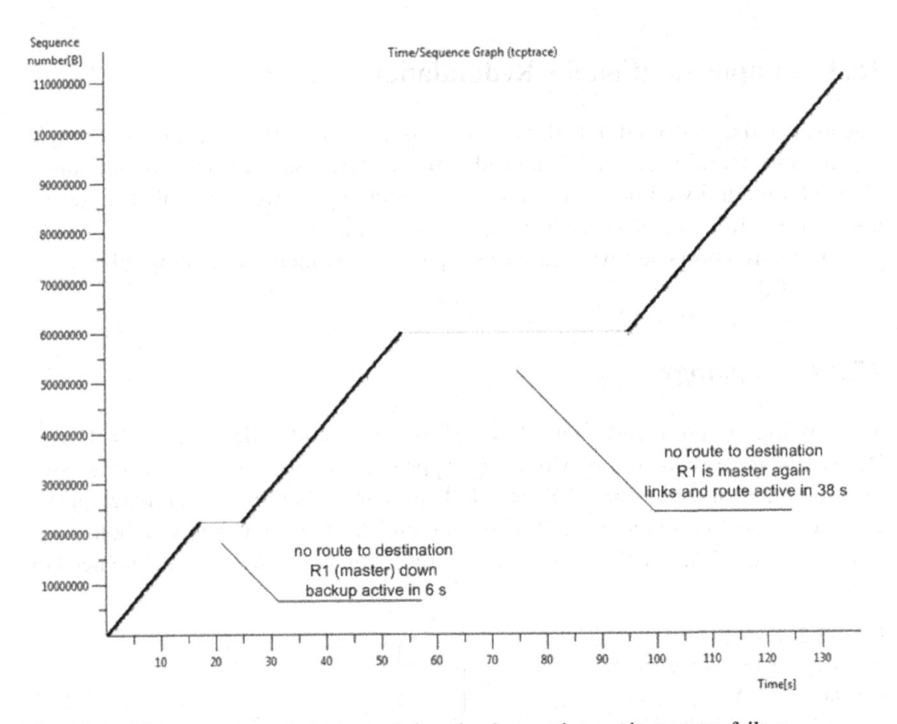

Fig. 15.3 TCP segment graph for typical download operation against router failure.

further reconfiguration of edge router - sometimes an important issue, as it may be administered by another party.

15.3.2 *Test Results*

Multiple tests were conducted with:

1. ICMP Connectivity (for break period measurement)
2. HTTP Download (for assessment of failures on quality of experience for typical web user)
3. Streaming (for assessment of failures on real-time communications)

15.3.2.1 Connectivity

The connectivity test consisted of "pinging" outside gateway for determining break period (time with no route to destination) after master router failure, and after bringing it on-line. Observations are presented on tab. 15.1.

Connectivity tests bring interesting observations - the break period is much bigger when putting master router back online. This is not a VRRP protocol outcome - if any of the switches in the infrastructure is using STP protocol (and it usually does), the STP convergence is the reason of such big values of break period. The solution would be using RSTP (but such switch would be more expensive) or not using STP at all.

15.3.2.2 Bulk Transfer

The bulk transfer tests were conducted to observe the effect of master router failure on typical bulk operations, suach as WWW browsing, downloading, etc. On fig. 15.3 one of representative cases of download rate and consecutive effects of router failure and repair is shown.

As expected, small break period during router failure has no effect on bulk transfer and quality of user experience. The bigger one, however, may lead to timeouts, depending on the applications configuration.

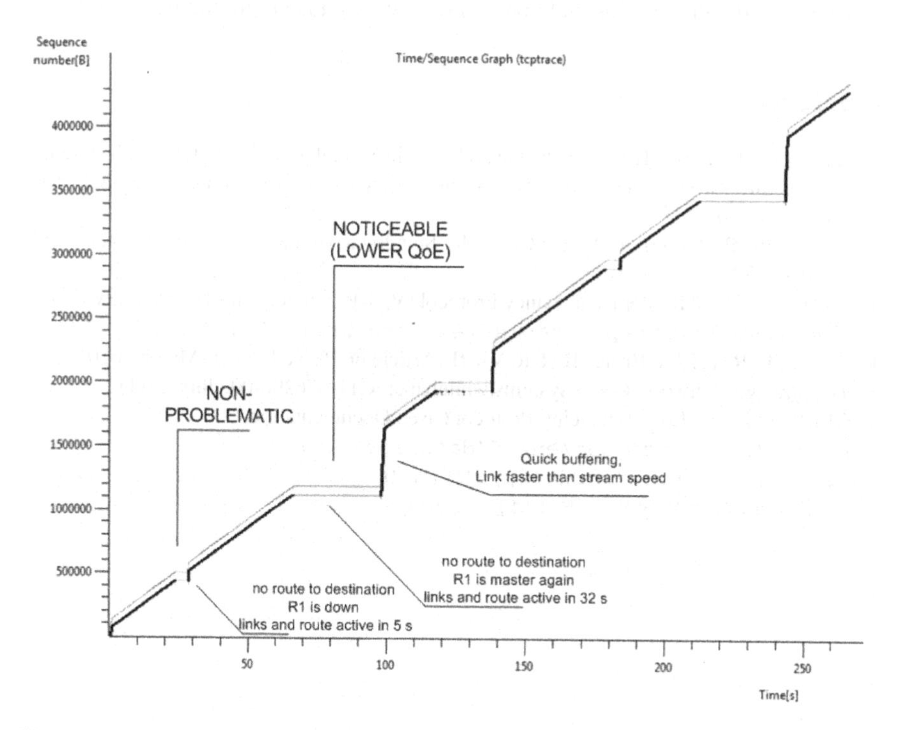

Fig. 15.4 TCP segment graph for typical streaming operation against router failure.

15.3.2.3 Streaming

As more and more businesses are using some kind of teleconferencing applications, and streaming video becomes major traffic element, tests of such application behaviour seem appropriate. Fig. 15.4 represents the sequence analysis of fixed-rate stream, which shows buffering and indicates needed buffer parameters. For example if a stream's rate is 62kbps, at least $40 * 62 = 2480kb = 310kB$ is necessary.

15.4 Conclusions

In this article, simple but effective routing redundancy solution is presented and tested. Redundancy is achieved with VRRP protocol utilization, and simplicity and cost-effectiveness requirements led to software routing solution Vyatta. Testbed for various situations was created and tests were conducted to determine how master router failure affects: (1) connectivity, (2) bulk transfer, (3) streaming traffic.

Presented solution is:

- Flexible and highly customizable (there can be more backup routers, and many services may be added, e.g. VPN or QoS funtions);
- IPv6 - ready (Both Vyatta and VRRP are IPv6 capable);
- Cost effective (Only single expense is the cost of two low-end servers and any switch, and configuration and maintenance workhours are minimal).

References

1. Garrett, B., Laliberte, B.: Brocade One and Hitachi Virtual Storage Platform. A Reference Architecture for Next Generation Data Centers. Enterprise Strategy Group Whitepapers (September 2010)
2. Li, T., Cole, B., Morton, P., Li, D.: Cisco Hot Standby Router Protocol (HSRP). RFC 2281 (March 1998)
3. Nadas, S.: Virtual Router Redundancy Protocol (VRRP) Version 3 for IPv4 and IPv6. RFC 5798 (March 2010), http://tools.ietf.org/html/rfc5798
4. Stretch, J.: RFC 5798 Brings IPv6 to VRRP. Article on Packetlife.net (March 2010)
5. The Strategic Network, Cisco Systems whitepaper, C11-676296-00 (July 2011)
6. GLBP - Gateway Load Balancing Protocol Cisco Documentation, http://www.cisco.com/en/US/docs/ios/12_2t/12/_2t15/feature/guide/ft_glbp.pdf
7. Vyatta.org Community Website, http://vyatta.org/

Chapter 16
The Example of IT System with Fault Tolerance in a Small Business Organization

Pawel Buchwald

Abstract. This chapter describes the relationship between technical aspects of IT infrastructure reliability and continuity of business activities in a small business organization. The paper presents a simple mathematical method of describing the relationship between requirements of business activity and reliability of IT infrastructure. This article shows the evaluation of the importance of IT services for organizations using the mathematical method presented in the example. This method allows the detection of bottlenecks and evaluate the effects of failures in the organization. The author selected IT services for which reliability must be improved, on the basis of mathematical analysis. This chapter presents methods of improving reliability of web services and databases using the failover Cluster and database mirroring.

16.1 An Analysis of IT Systems Reliability In Order to Evaluate the Continuity of Business Services

High availability of IT systems can ensure correct functionality of business organizations. Today business processes realized in organizations depend on IT infrastructure. Long breaks in normal operation of IT equipment and software can't be accepted. Organizations providing services which determine the human health or business safety require highly availability IT infrastructure. Availability can be defined as the ratio of regular work time and the system's lifetime.

Downtime is the time required to resume services after a shutdown or a failure. In the normal exploitation of IT infrastructure administrators must attempt to take maintanance operations such as patch installation or hardware replacement. In many

Pawel Buchwald
Academy of Business in Dabrowa Gornicza, Department of Computer Science
ul. Cieplaka 1c, 41-300 Dabrowa Gornicza, Poland
e-mail: pawel@buchwald.com.pl
http://www.wsb.edu.pl

A. Kapczynski et al. (Eds.): Internet - Technical Develop. & Appli. 2, AISC 118, pp. 179–187.
springerlink.com

$$A = \frac{T1\ work}{T1\ work + T2\ downtime}$$

Fig. 16.1 Formula to calculate availability

cases, such tasks require the system shutdown. Availability of IT infrastructure with fault tolerance is 99.999, which means that an average downtime is 5 minutes in a year including maintenance tasks. High availability is usually achieved by means of redundancy. In architecture with redundancy many homogeneous devices work as a one device. Downtime of a one device will activate the next redundant equipment. The probability of a failure in such a system can be determined as multiplying the probabilities of a failure of the whole redundant equipment.

$$P = \prod_{i=1}^{n} Pi$$

Fig. 16.2 Formula to calculate the probability of a failure

The probability of the trouble-free operation can be calculated as

$$P = 1 - \prod_{i=1}^{n} Pi$$

Fig. 16.3 Formula to calculate the probability of the trouble-free operation

n - no. of redundant equipment Pi - Probability of failure single equipment

The probability of failure single equipment can be given a priori by a manufacturer's data or can be determined experimentally. A large number of redundant equipment increases the cost of the IT system. The budget for the IT infrastructure is limited, so the number of additional devices can not be too high. Redundancy is used above all for devices that support the operation of key business processes. The table 1 shows the relationship between the availability and the amount of redundant equipment. The example assumes that the probability of a failure of a single device is 0.05. Using four redundant devices ensures the availability which is characteristic for systems with fault tolerance.

A Selection of devices that are to be redundant in the first place must satisfy requirements of key business processes. The device selection for redundancy may be made by using dependency matrix between the devices and business processes in an organization. The Dependency matrix contains information about using the device to perform business operations or processes. For each of the processes can be

count of devices	probability of fault	Availability
1	5,0000000000%	95,0000000000%
2	0,2500000000%	99,7500000000%
3	0,0125000000%	99,9875000000%
4	0,0006250000%	99,9993750000%
5	0,0000312500%	99,9999687500%
6	0,0000015625%	99,9999984375%
7	0,0000000781%	99,9999999219%

Fig. 16.4 Impact between number of redundant device and availability

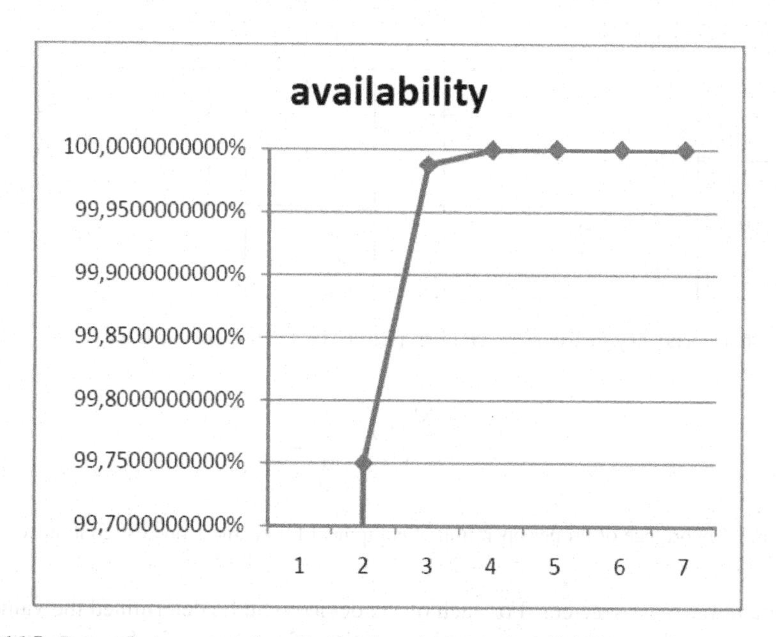

Fig. 16.5 Impact between number of redundant device and availability

specified the number representing the priority of a process in the organization. A high priority means higher reliability requirements. An Example of a dependency matrix is shown in Fig. 15.6. Each device in dependency matrix has an availability factor, which determines the device reliability. The higher value of this factor means greater reliability. Matrix elements determine whether a given business process is dependent on the correct work of the device. Value ‚1' means the relationship between the business process and equipment, while the value 0 means a lack of dependency. Dependency matrix assess the impact of device failure on conditioning of

	Priori ty	1.00	0.70	0.70	0.60	0.50	1.00	0.90
availability factor	Business processes	sending orders	receiving emails	browsing website	monitoring of Workers	downloading a form for employees	processing orders	quality control
97,00%	WWW Server	1	0	1	0	0	1	0
97,40%	Database Server	1	0	1	1	0	1	1
99,90%	Router	1	1	1	1	0	0	0
99,10%	Switch 1	1	1	1	1	1	1	1
99,12%	Switch 2	0	0	1	1	1	0	1
96,00%	FTP Server	0	0	0	0	1	0	0
98,00%	SMTP/POP3 Server	0	1	0	0	0	0	0

Fig. 16.6 Dependency matrix between IT infrastructure and business process

$$I_i = \frac{\sum_{j=1}^{n} P_j M_{ji}}{\sum_{j=1}^{n} P_j}$$

Fig. 16.7 Importance of proper operation of equipment for business process continuity

important business services. For each of the devices can be determined the value I, which defines the importance of proper operation of equipment for business process continuity. where Ii - value of importance factor I for device number i n - number of business process Pj- priority of business process no. j

The dependency matrix can also be useful for determining the relationship of availability of a business process or activity. The Availability of a business process is not greater than the availability of the most unreliable equipment which is necessary to execute the process. Business process availability factor Bi can be calculated from the formula: Bi=min {Mji *Ai : i=1..n} where Mji- element ji of dependency matrix; Ai - availability of device no. i; Bi- availability of business process no. i

To determine the factor Bi and factor Ii, it is necessary to know the priorities of the business processes. In most cases, the knowledge of business processes is not

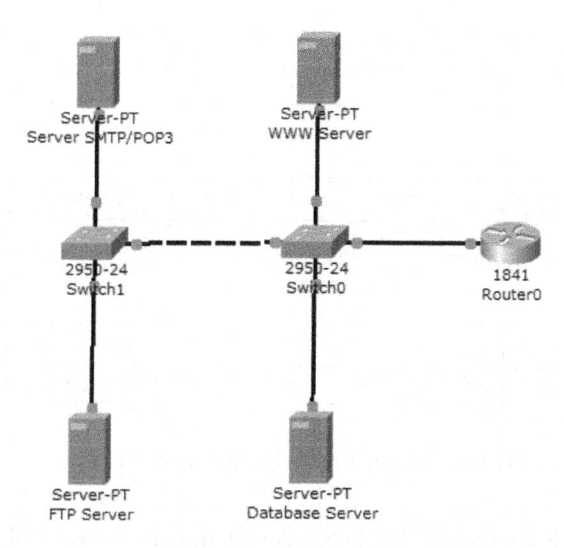

Fig. 16.8 IT infrastructure of a small business organisation

known by IT specialists. Information about the priority of the given processes must be specified by specialists of business management. The priorities of the processes may change over time, so the factors Bi and li should be periodically recalculated.

The analysis presented in the example can be used for the detection of key equipment. Technical support personnel should specifically ensure the availability of such devices and to minimize their downtime. The detection of key devices is useful in the analysis of bottlenecks. Devices used in many processes can be also overexploited. For this reason, redundancy methods are used together with load balancing. The Knowledge of key devices and their loads is fundamental to choose the method of technical infrastructure redundancy.

The matrix in fig. 15.6, which was presented in the example, relates to the technical infrastructure shown in Fig 8. We can calculate the factors of importance of each exhibited devices. For business processes we can calculate the value of availability of each business process. The value of availability depends on the reliability of the bottleneck (device with the worst reliability). Devices, for which the value of I is above the average, are important for maintaining business continuity. The execution of many important services of the organization will not be possible, if one of these devices breaks down. The most important processes in the organization should be highly reliable. For this reason, one should improve reliability of the equipment needed to perform the most important processes. In particular, the reliability of the device, which is the bottleneck of the major business processes should be improved.

The value of factor I calculated for each devices is shown in Fig. 15.9. Table exhibit on fig. 15.9 presents equipment, which is a bottleneck of important business processes.

Switch 1 and database server are two devices having the highest value of Importance Factor. WWW server and database server are bottlenecks for many business

Device Name	Importance Factor	Bottleneck of processes or events
WWW server	0,500	sending orders, browse website, processing orders
Database Server	0,778	monitoring, quality controlling
Router	0,556	-
Switch 1	1,000	-
Switch 2	0,500	-
FTP Server	0,093	downloading forms
SMTP/POP3 Server	0,130	receive emails

Fig. 16.9 Importance factors for equipment and bottlenecks for process

processes. The Reliability of dedicated network devices, shown in the example, is above the average. The reliability of servers is too low and needs improvement. The administrator should especially increase the reliability of the database server because it has a greater impact on the organization's business continuity processes. Improving the reliability of the web server, which is a bottleneck, will reduce the risk of interruption of many business continuity processes and events. For this reason, we should improve the reliability of these two servers.

16.2 Selected Mechanisms of Redundancy Used to Improve the Reliability

SQL Server database mirroring feature, offers new functionality that allows you to configure database failover. Database mirroring is primarily a software solution for increasing database availability. Mirroring is implemented on a per-database basis and works only with databases that use the full recovery model. Database mirroring maintains two copies of a single database that must reside on different server instances. One server instance hosts the database to clients (the principal server). The other instance acts as a hot or warm standby server (the mirror server).

Database mirroring involves redoing every insert, update, and delete operation that occurs on the principal database onto the mirror database as quickly as possible. Redoing is accomplished by sending a stream of active transaction log records to the mirror server, which applies log records to the mirror database in the given sequence, as quickly as possible. Unlike replication, which works at the logical level, database mirroring works at the level of the physical log record. Both principal and mirror roles are typically interchangeable in a process known as role switching. Role switching involves transferring the principal role to the mirror server. In role switching, the mirror server acts as the failover partner for the principal server. When a role switch occurs, the mirror server takes over the principal role and brings its

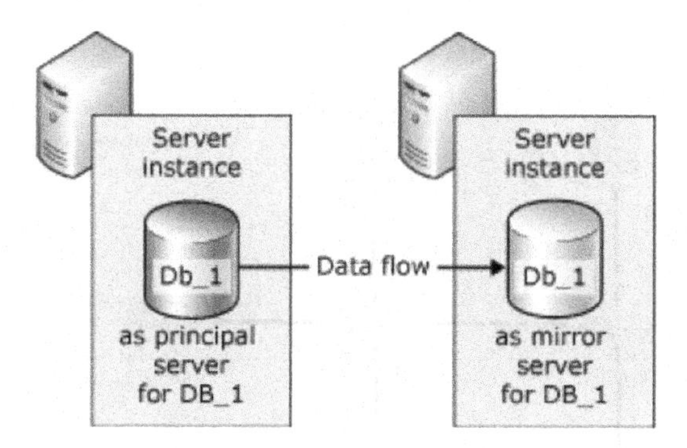

Fig. 16.10 Mirroring mechanism in SQL Server

copy of the database online as the new principal database. Working of mirroring mechanism is shown in Fig. 15.10. The witness server role is an optional role used to configure database mirroring. It is used to detect any failures and failover. It can be configured by using the high availability operating mode. The witness server serves multiple mirroring pairs. Mirroring pair includes one primary and one mirror database. If the primary database fails and the failure is confirmed by the witness server, the mirror database takes the primary role to serve the users of the database.

Redundant Web servers are two or more Windows computers that provide website services for company network. A redundant Web server acts as a failover computer, which means that if primary Web server fails, the secondary failover server takes over. This means there is no downtime for WWW service. Microsoft Windows Server operating system supports Failover Cluster Management. This functionality enables redundancy of web server, which helps to increase the reliability.

The Reliability of Web servers and database server can be improved by using failover cluster and database mirroring. Web server can host database instance, which will play the role of the mirror server for database service in exhibit infrastructure. The machine that hosts database principal server can be also a failover node for web services. The new location of web services and database services in company's LAN is shown in Fig. 15.11. In new configuration the probability of services failure is in fact the probability of failure of the two machines at the same time, which are used to host web services and database instances. The probability of fault is equal:

$$P = (1 - 0{,}97) * (1 - 0{,}974) = 0{,}00078$$

The probability of fail-free operation can be calculated as

$$1 - P = 1 - 0{,}00078 = 99{,}92$$

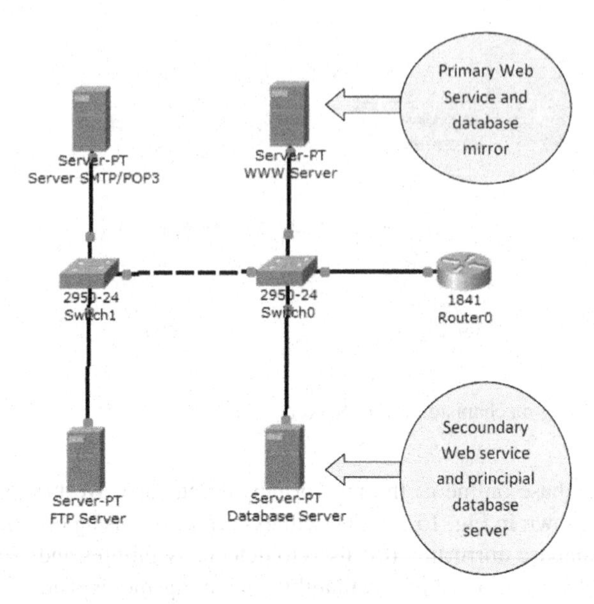

Fig. 16.11 Location of redundant services

This can be considered as a new coefficient of reliability for web services and databases in the organisation. The use of redundant mechanisms by the introduction of database mirroring and web failover clustering improves the reliability of services.

16.3 Conclusions

The Redundancy methods can improve the IT service reliability. Improving efficiency can be realized by redundancy of equipment and services. Redundancy, using the software methods like database mirroring or failover clustering does not require additional physical devices. It reduces economic cost of solutions. Using the reorganization of IT services without replacing equipment can increase the reliability, as shown in the example. The diagnosis of reliability requirements needs to be made taking into account requirements of the organization's business processes. The relationship between the equipment and business activities in an organization can be appointed by using mathematical methods. The mathematical description of the relationships between business processes and technical equipment can help in assessing the effects of the accident for the organization. dependency matrix, presented in the example can be used as a simple description of relationship between IT services and business requirements. This method is simple and gives the possibility of using mathematical formulas understandable for IT professionals and business professionals. The requirements of business processes may change the course of time. Depending on business needs, analysis and reliability of IT services must be performed periodically. A proper analysis of business needs and the use of infrastructure with fault tolerance can increase the reliability of service companies, as shown in this chapter.

References

1. Grzywak, A. (ed.): Bezpieczenstwo Systemow Komputerowych; Wydawnictwo Pracowni Komputerowej Jacka Skalmierskiego (2001)
2. Stanek, W.: Microsoft SQL Server 2008 Administrator's Pocket Consultant. Microsoft (2008)
3. Michael, L., Mansfield, M.: SQL Server 2008 Administration Instant Reference. Wiley John + Sons (2009)
4. Patel, K., Snyder, W., Knight, B.: Professional Microsoft SQL Server 2008 Administration. Wrox (2008)
5. Stern, H., Marcus, E.: Blueprints for High Availability. Wiley John + Sons (2003)
6. Tulluch, M.: Introducing Windows Server 2008. Microsoft Gmbh (2007)
7. Johanson, M.: Windows Server 2008 Security. Resource Kit. APN Promise (2008)
8. Doherty, J., Anderson, N., Della Maggiora, P.: Cisco Networking Simplified. Cisco Systems (2008)

Chapter 17
Local and Remote File Inclusion

Michal Hubczyk, Adam Domanski, and Joanna Domanska

Abstract. In this paper we have examined Local and Remote File Inclusion vulnerabilities in details. It is proven that these security flaws may lead to variety of problems including generation of big load to the server as well as disclosure of files which should not be accessible by the clients. These attacks are not so popular as SQL injection or Cross-Site Scripting but still they are subject of serious threat because of their consequences.

17.1 Introduction

Security of websites is very important since the first dynamic web pages [1] were presented. Nowadays web pages tend to act like applications with some user interaction, so the visible content and overall behavior of website may change due to action of user. However this may lead to serious consequences, because web developers [2] more often are working on increasing the usability and functionality than on security problems. From the authors observations and experience in web development this is caused by both: web developer unawareness about security flaws created (and left) in process of preparing web applications as well as time limitation problems. While source of the first cause is rather obvious, because quality of code strongly depends on developer's experience, than second case is caused by the fact that security issues

Michal Hubczyk · Adam Domanski
Institute of Informatics
Silesian Technical University
Akademicka 16, 44–100 Gliwice, Poland
e-mail: adamd@polsl.pl

Joanna Domanska
Institute of Theoretical and Applied Informatics
Polish Academy of Sciences
Baltycka 5, 44–100 Gliwice, Poland
e-mail: joanna@iitis.gliwice.pl

A. Kapczynski et al. (Eds.): Internet - Technical Develop. & Appli. 2, AISC 118, pp. 189–200.
springerlink.com

are neither functional, nor visual part of the web application, so the customer who is paying for the developer's work most likely will not pay for additional time of the developer to make the website more secure.

The main reason of vulnerabilities in dynamic web pages is non-proper validation of user input. Web developers often treat web user interfaces only as the source of user input, which is not a valid axiom.

The requests to run dynamic scripts do not have to come from the ??? (dynamic) website itself. Furthermore, the attacker (person who is trying to exploit the website functionality) may be able to use his own tools to automate the process of attack. The results of such actions may vary depending on the type of attack and the website structure. However, if the attack is successful the intruder may be able to compromise whole system which is hosting hundreds of websites just by using one vulnerable web application. There are many classifications of web application vulnerabilities, however, we propose our own classification which is consistent and clear-depending on target of the attack. We may distinguish three main types of web application vulnerabilities:

- Web application vulnerabilities — those which are present in code of the web application.
- System vulnerabilities — those which are present in the system which is hosting the web application.
- Mixed vulnerabilities — those which depend both on the application and system. The problem is in non proper configuration of the server and in web application.

Furthermore we divide Web application vulnerabilities on four types:

- Client-side vulnerabilities — those which are present on the client-side of the application, so vulnerable part of the application runs in client environment
- Server-side vulnerabilities — those which are present on the server-side of the application, so vulnerable part of the application runs in server environment
- Mixed vulnerabilities — those which depends both on the client, and server-side of the application
- Miscellaneous vulnerabilities — those which could not be assigned to the rest, however they are still important in terms of web application security

In this work we focus on one type of vulnerabilities which is dangerous and popular nowadays i.e. File Inclusion which has two variants:

- Local File Inclusion (LFI)
- Remote File Inclusion (RFI)

Local and Remote File Inclusion vulnerabilities are classified by the author as Web application vulnerabilities — Server-side vulnerabilities, because they are dependent on the scripts which are executed in the server environment.

The Imperva company has released report [3] in which it claims that RFI attacks have rather low volume: „*The relative volume of RFI attacks is usually low. For sake of comparison, we measured the frequency of observed RFI attacks against the frequency of observed SQL Injection attacks between December 2010 and March*

2011. Within the attack traffic, 1.7% was associated with SQL Injection, while 0.3% of the attack traffic was identified as RFI-related"

The relative percent values are rather low, however values that can be found in report are impressive in terms of web application security: from 30 November 2010 to 14 April 2011 there was about 10 attacks per minute detected, however there are peaks of attacks oscillating even at value of 425 attacks per minute, which in scale of the year gives rather big number. The LFI and RFI vulnerabilities cause information disclosure to the attacker which may lead even to destroying of the webpage. However we have noticed that special case of the LFI attack leads to increased server memory consumption and processors usage - the script may enter into endless loop which is caused by recursive self inclusion of the script. This may lead even to Denial of Service attack [4].

17.2 Local File Inclusion and Remote File Inclusion

The inclusion of other files is very common in PHP scripts, for example the realization of including classes or functions for future use may be implemented using include() [5], require() [6], include_once() [7], require_once() [8] PHP functions. These functions take one parameter which is path and filename of the file that should be read and parsed by the PHP interpreter. The main difference between Local and Remote File Inclusion vulnerability is that in first case the included files are stored on the local server from website point of view. In latter case the file may be included from the remote location e.g. from the attacker's server.

17.2.1 Explanation of the Vulnerability

There are some possible problems with processing the included files. Some websites provide navigation by including the file according to parameter stored in the URL [[9] for example: *http://example.com/index.php?site=home.html*. There is possibility then that the „site" variable is name of the file which is to be included, so the attacker could try to manipulate the variable's value to exploit the functionality of the website, or even get data which he should not have access to. Such case is rather obvious and is not popular nowadays. However the extension of the file („.html in this case") may be added to the parameter in the script. The web developer may not know that the attacker could access non-html files. There are methods to bypass protection mechanisms, by adding e.g. „%00" which is Null character [10] that terminates the string in PHP parser in some environments.

17.2.2 Example 1

This example covers the simplest case of Local File Inclusion vulnerability. The parameter „page" takes the name of file that should be included. In this case the web interface offers two pages to load: "pages/home.html" and „pages/other.htm".

```
<?php
if(isset($ _GET['page']))
   {
      $ content = "pages/".$ _GET['page'];
   }
else
   {
      $ content = "pages/home.html";
   }
?>
<div>
<a href="1.php?page=home.html">Home page</a>
<a href="1.php?page=other.html">Other page</a>
</div>
<div>
</div>
<div style="height: 300px; background: yellow;
padding: 10px">
<?php include(\$ content) ?>
</div>
```

Script 1. Application used to present Local File Inclusion vulnerability

The attacker is able to change „page" parameter and try to retrieve other files than developer expects. According to Script 1 Application used to present Local File Inclusion vulnerability the script does not validate the input, so it is possible to perform the attack.

17.2.2.1 Attack Scenario 1 — Generating Server Load

The attacker tries to input random string instead of file name into the „page" parameter to check if the script filters somehow the input.

The attacker enters such URL in web browser address bar (exemplary address):

```
http://localhost/mgr/examples/serverside/
lfirfi/1.php?page=random_string
```

The exemplary response:

```
Warning: include(pages/random\_string)
[function.include]:

failed to open stream:
No such file or directory in
C:/wamp/www/mgr/examples/serverside/
lfirfi/1.php on line 18
```

```
Warning: include() [function.include]: Failed opening
'pages/random\_string' for inclusion
(include_path='.;c:/wamp/bin/php/php5.3.3/pear') in
C:/wamp/www/mgr/examples/serverside/
lfirfi/1.php on line 18
```

Above simple experiment gave the attacker very important information, because he now knows how dynamic script works — it uses include() [5] function to load a file from parameter „page" in „pages" directory. Now attacker may try to read some other files on the remote server. The intruder decided to see what will happen if he input „../1.php" as the „page" parameter's value. The script does not validate the extension, and directory traversal is not blocked (using „../" to go to parent directory), so the attack was somehow successful. The script has started endless loop, because each inclusion of „1.php" was causing another self inclusion until the stop criterion was reached (memory limit in this case-set to 528MB, however maximum execution time would also stop the script if it was first exceeded). To draw conclusions of how the attacker would affect the server, the system monitor was examined while the attack was performed. Figure 17.1 shows the memory usage curves while the script was executed, we can distinguish execution times by looking inside red circles marking the process of memory allocation by the server. From authors observations the script has ended its execution while httpd.exe process has reached about

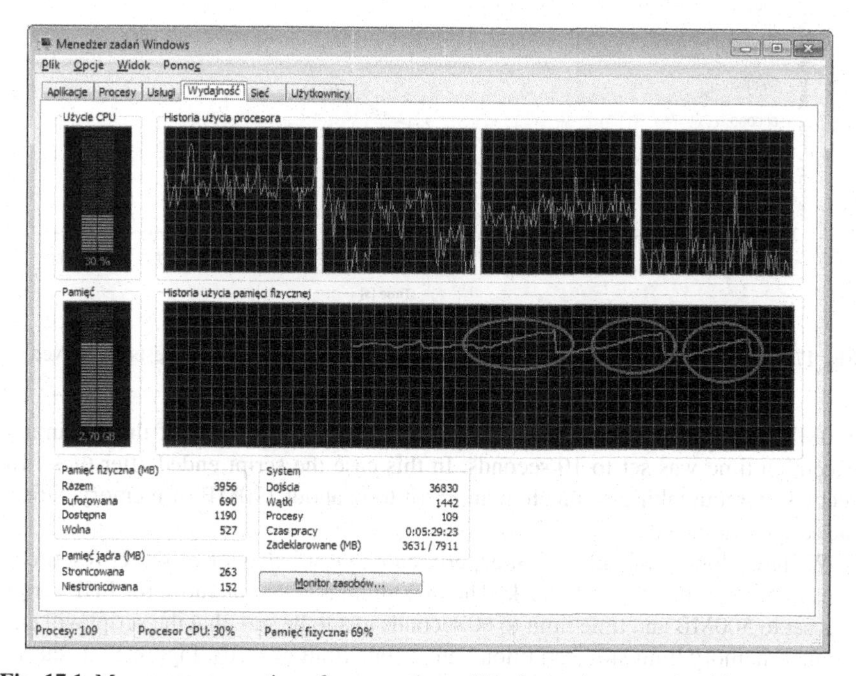

Fig. 17.1 Memory consumption after executing self inclusion loop attack with use of Local File Injection vulnerability

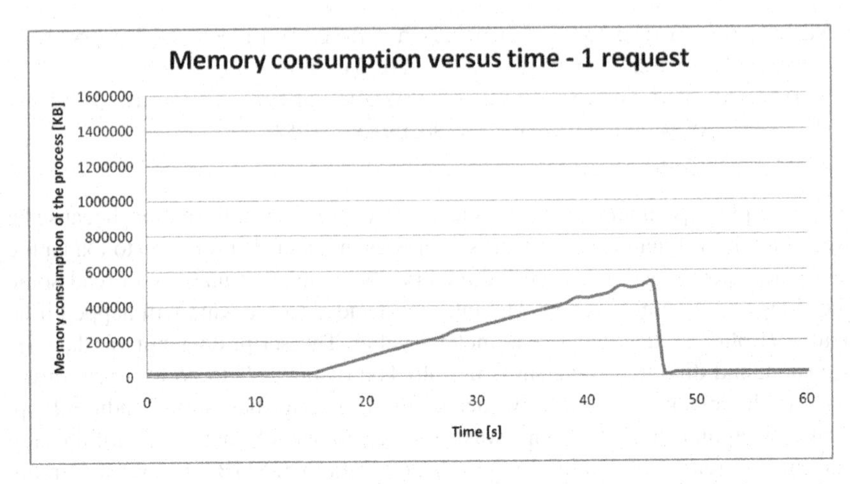

Fig. 17.2 Memory consumption of the HTTP servers process in time-1 request is served

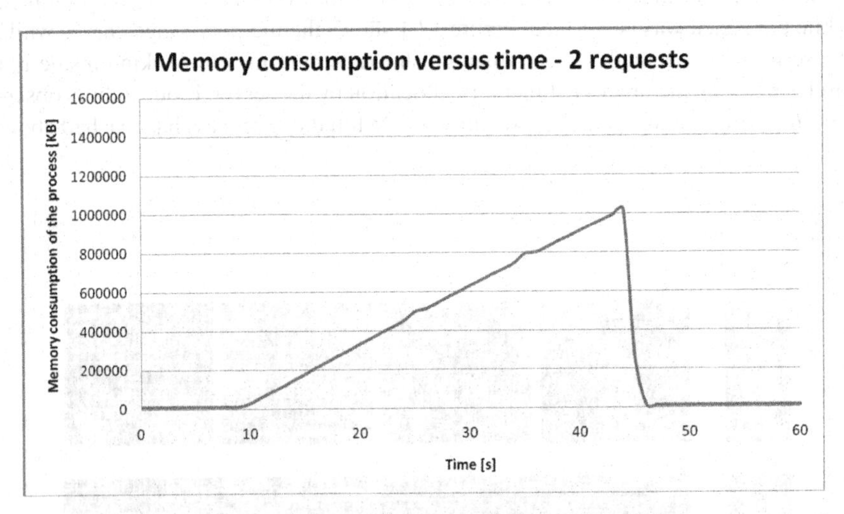

Fig. 17.3 Memory consumption of the HTTP server process in time-2 requests are served

540MB of memory. The attacker also checked what will happen if the maximum execution time was set to 10 seconds. In this case the script ended after those 10 seconds without taking so much memory (it took about 190MB of memory before the script terminated).

We have done some additional experiments to test behavior of server resources consumption in this type of attack. The maximum allowed memory for PHP script was set to 500MB and time limit to 60 seconds just to be sure that the script will encounter memory limit stop condition before time limit exceeds. Figure 17.2. shows the memory consumption of the HTTP server process while it processes only one request. The curve reaches maximum (0.5GB of memory) in about 30 seconds, then

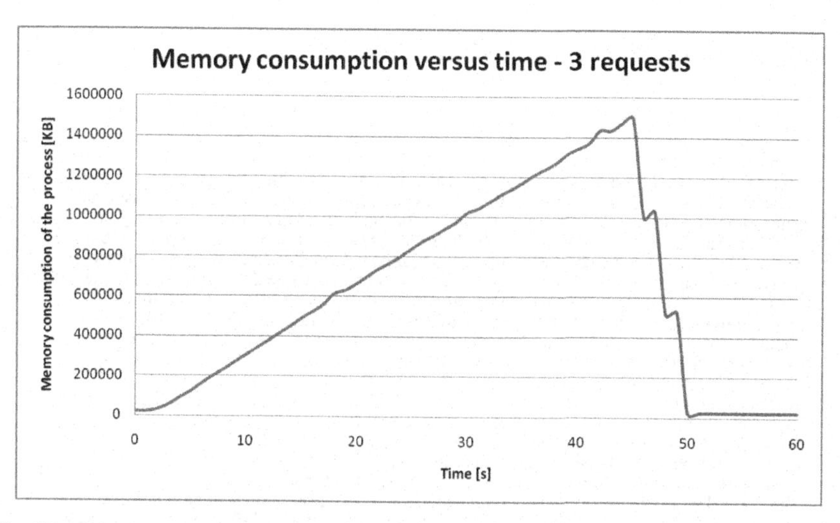

Fig. 17.4 Memory consumption of the HTTP server process in time-3 requests are served

it is going down (moment of the script termination). Figure 17.3 shows the same, but for two simultaneous requests, the curve is a bit steeper and it reaches maximum (about 1GB of memory) in about 35 seconds. The last examined case is shown in the Figure 17.4, where three simultaneous requests are processed. It reaches its maximum (about 1.5GB) in about 50 seconds. The curve is steeper, so the speed of memory allocation increases with number of requests, thus the processors usage should be greater. It can be noticed that curves have linear tendency in time, so if the limit of memory was greater, then the script would take proportionally more time to reach the maximum.

The processor usage was very similar to the memory consumption behavior. The tests were made using a computer with Intel i5-560M processor which has 2 cores with Hyper-Threading technology [11]. In the first case (1 request) the HTTP server process took 25% of processor time, in the second case (2 requests) it took about 48%-50% of processor time. In the last case (3 requests) it took about 60%-70% of processor time. It is clearly visible that with increase of concurrent requests, the processor usage increases. Even if the maximum execution time and the memory limit were set to relative low values, then the attacker would just need to increase amount of concurrent requests to achieve similar results.

17.2.2.2 Attack Scenario 2 — Reading Not Application Related Files

The attacker now knows that he is able to change argument of include() [5] function, so he has decided to get more information. The very interesting file in Unix [12] and Unix- like [13] operating systems is „/etc/passwd" which stores information about accounts in the system. If the page is hosted on such operating system and the user assigned to HTTP server is able of reading „/etc/passwd" file, then the attacker may

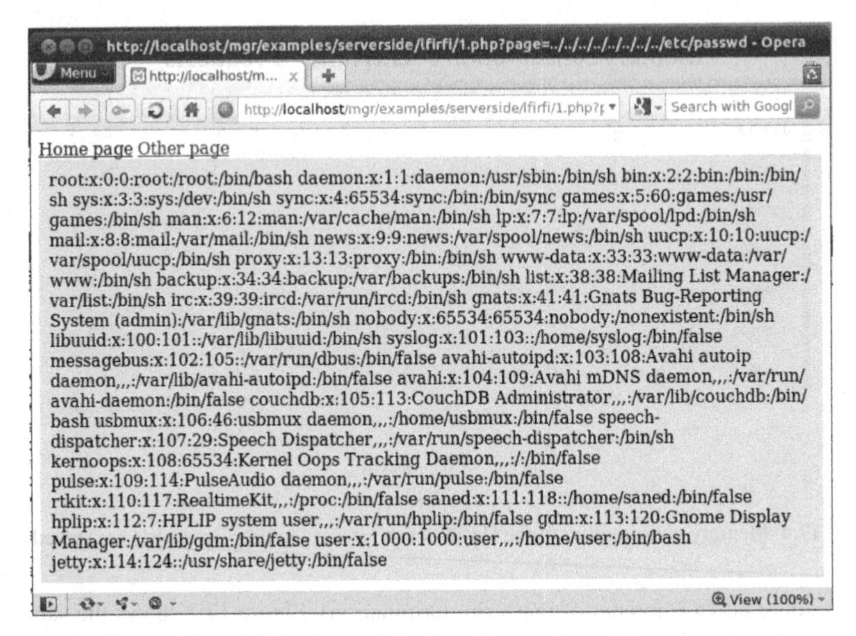

Fig. 17.5 Successful read of /etc/passwd file from the remote machine

be able to embed the file into the vulnerable webpage to reveal accounts information on the server.

The intruder probably already knows the absolute path to the web application from the first example, because PHP will most likely print warning message if include() [5] function will not find file from its argument. However even if error reporting is turned off in the PHP configuration, then the attacker could use basic knowledge about file system to guess that „passwd" file should be present in „/etc" directory, so it is easy to access it by adding additional „../" to the function argument, to traverse directories. In this case the attacker have set „page" parameter to such value: „../../../../../../../../etc/ passwd". This means that he has traversed through eight directories. The Figure 17.5 shows the results of successful Local File Inclusion attack in which the attacker got the contents of „/etc/passwd" file. Theoretically the intruder now may have access also to other system files if the read rights are sufficient.

17.2.3 Example 2

This example is very similar to previous one, but covers possibility of including remote file. This is even more dangerous, because the attacker could use his own scripts and execute them on the remote location.

The script is very similar as in previous example, but it searches for file in the current directory (see Script 2 - PHP script used to present Local and Remote File Inclusion attack). Navigation is done by including the file which name is taken from the „page" parameter in URL.

```
<?php
if(isset($_GET['page']))
   {
      $content = $_GET['page'];
   }
else
   {
      $content = "home.html";
   }
?>
<div>
<a href="1a.php?page=home.html">Home page</a>
<a href="1a.php?page=about.html">About</a>
</div>
<div>
</div>
<div style="height: 300px; background: #dedede;
padding: 10px">
<?php
   include(dollar_sign_content);
?>
</div>
```

Script 2. PHP script used to present Local and Remote File Inclusion attack

When navigating the website through links in the upper menu, the website looks as it should look like, because it is how the web developer has designed it. However if the intruder tries to input address of the website on the remote server by means of HTTP protocol, then vulnerable webpage behaves unexpectedly. The user input was not filtered properly and additionally server configuration allowed fetching file from the remote server. In this case the attacker set address of Silesian University of Technology (http://www.polsl.pl) as a value of the „page" parameter. On the screen website of Silesian University of Technology is visible in the gray field (see Figure 17.6). That means that the website has critical vulnerability of Remote File Inclusion.

This vulnerability may lead to serious problems, because the attacker could point the script to include his malicious code. If the system functions are enabled in the website environment, then he could use scripts written in PHP that mimic the shell behavior of the operating system. It is very easy to write such script, the authors wrote the very basic example of PHP shell in less than 10 lines (see Script 3).

```
<form action="<?php $_SERVER['PHP_SELF']?>" method="post" >
<input type="text" name="cmd" />
<input type="submit" value="enter" />
</form>
<?php if(isset($_POST['cmd']) AND $_POST['cmd'] != NULL):?>
<b>Last command: <?php echo $_POST['cmd'] ?></b><Br/>
<pre><?php system($_POST['cmd']);?></pre>
<?php endif ?>
```

Script 3. PHP Example of simple PHP shell

The script utilizes system() [14] function to execute operating system commands and print the output to the screen. Such PHP shell could be included by the intruder instead of e.g. Silesian University of Technology website address. To make such script to work in the remote environment, the server which is delivering PHP shell cannot parse PHP scripts, because vulnerable website would get only the resulting code of the web application. It is because HTTP server with PHP support is usually configured to parse „.php" files, so the users get parsed code instead of source code of the script. On the other side it is not so hard to find or even create such server — it is even possible for the attacker to disable the PHP parser temporarily just by switching one option in php.ini configuration file.

Include() [5] function parses content of the file which it is processing, so if it encounters PHP script, then the PHP parser would parse it locally and output results. The threat of processing malicious PHP scripts is very dangerous, because depending on configuration of the server — the attacker can perform more or less destructing actions. On the other hand, he may steal some sensitive data without leaving much information about his „visit" For example intruder may read the contents of „index.php" files of web applications (e.g. „cat" in Linux and „type" in Windows

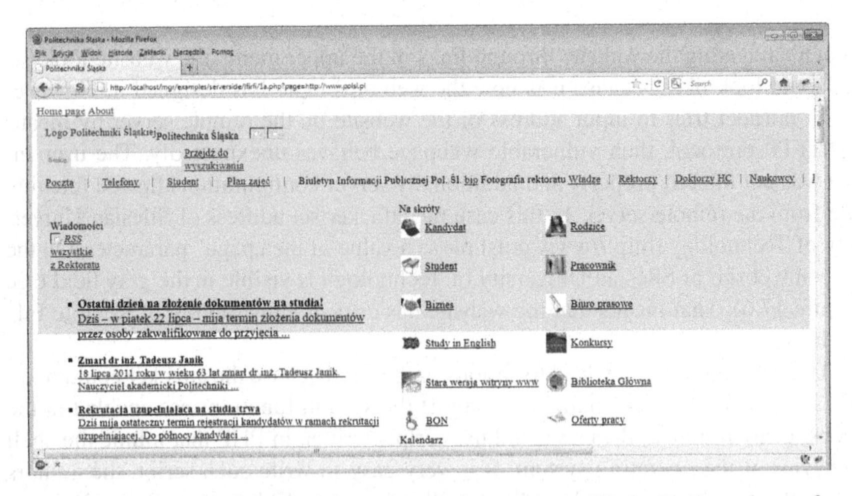

Fig. 17.6 Unexpected behavior of the webpage, the content is filled with the webpage from external server after successful Remote File Inclusion attack

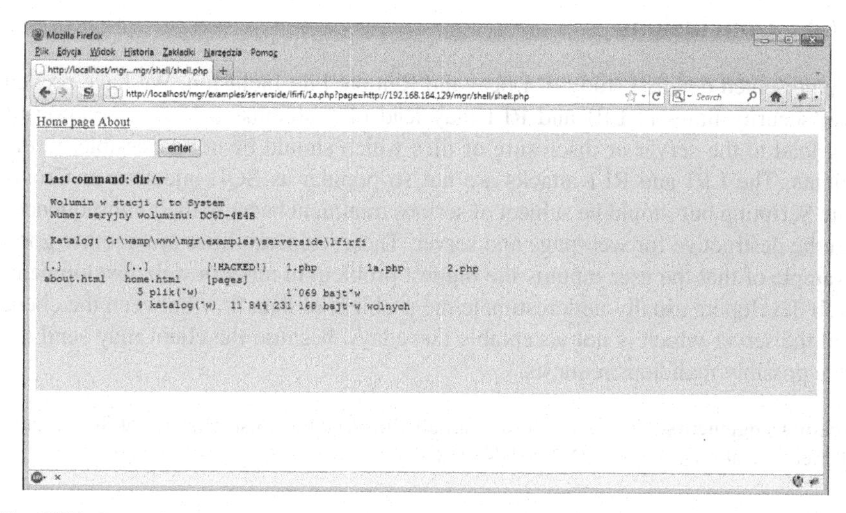

Fig. 17.7 Example of Remote File Include vulnerability with injected PHP shell script.

environments) that are hosted on the vulnerable server, from where the database connection strings may be read. If the attacker has sufficient rights, he may even destroy the webpage, or inject some Cross-Site Scripting [15] codes on the hosted websites to steal cookies from the visitors. Example of such attack on website hosted on Windows 7 operating system is visible in the figure 17.7. The command input field was first filled with „mkdir !HACKED!" which creates new directory and then with „dir /w" phrase which does directory listing. Below is „Last command:" text which is followed by last inserted phrase in the command field. Then the results of executed commands are printed. Now the attacker has shell access to the server with rights of user assigned to HTTP server process.

17.3 How to Protect

Protection from Local File Inclusion and Remote File Inclusion vulnerabilities is in fact not sophisticated, but web developers should keep in mind that the attacker could find new way of bypassing the security checks. However the author is able to give some rules of thumb which for sure will help with the problem. First of all it is important to filter argument of include() [5] and related functions if there is even small part on which client has control. If the web developer finds „://" string in function argument, then he can be sure that someone tries RFI attack on his website. To protect from LFI, web developer can prepare list of available files to include and check if included file is on the list. More dynamic way of dealing with this problem would be filtering the „../" and „./" strings which would suggest try of directory traversal.

17.4 Conclusions

We presented detailed study of Local and Remote File Inclusion. We have proved that security flaws in LRI and RLI may lead of problems such as generation of big load to the server or disclosure of files which should be not accessible by the clients. The LRI and RLI attacks are not so popular as SQL injection or Cross-Site Scripting but should be subject of serious treatment because their consequences can be destructive for web page and server. These vulnerabilities are another good example of that the user input is the biggest problem in modern web development. Web developers usually underestimate the problem of data flow between the client and the server which is not acceptable nowadays, because the client may send his own possibly malicious requests.

Acknowledgements. This research was partially financed by Polish Ministry of Science and Higher Education project No. N N516441438.

References

1. http://en.wikipedia.org/wiki/Dynamic_web_page (July 22, 2011)
2. http://en.wikipedia.org/wiki/Web_developer (July 22, 2011)
3. Imperva company, Hacker Intelligence Initiative, Monthly Trend Report #1
4. http://en.wikipedia.org/wiki/Denial-of-service_attack (July 23, 2011)
5. http://php.net/manual/en/function.include.php (July 23, 2011)
6. http://php.net/manual/en/function.require.php (July 23, 2011)
7. http://www.php.net/manual/en/function.include-once.php (July 23, 2011)
8. http://www.php.net/manual/en/function.require-once.php (July 23, 2011)
9. http://en.wikipedia.org/wiki/Uniform_Resource_Locator (July 23, 2011)
10. http://en.wikipedia.org/wiki/Null_character (July 23, 2011)
11. http://www.intel.com/technology/platform-technology/hyper-threading/index.htm (July 24, 2011)
12. http://en.wikipedia.org/wiki/Unix (July 24, 2011)
13. http://en.wikipedia.org/wiki/Unix-like (July 24, 2011)
14. http://php.net/manual/en/function.system.php (July 24, 2011)
15. http://en.wikipedia.org/wiki/Cross-site_scripting (July 24, 2011)
16. https://www.owasp.org/index.php/SQL_Injection (July 24, 2011)

Chapter 18
Maildiskfs - The Linux File System Based on the E-mails

Adam Domanski and Joanna Domanska

Abstract. This article describes the Maildiskfs - an efficient, convenient and fully secure data storage mechanism holding the data on the mail servers. This implementation is a new approach of the storing data problem. This article also describes the performed test of the system performance based on a comparison with other similar solutions. Maildiskfs project shows that it is possible to create convenient data storage system using the e-mail mechanisms.

18.1 Introduction

This article describes the Maildiskfs - an efficient, convenient and fully secure data storage mechanism holding the data on the mail servers.

The Maildiskfs is a file system which works similar to the commonly used systems, but it has the storage mechanisms which works quite different. This system is designed only for the GNU/Linux operating system. Construction and operation of file system mechanisms significantly varies between different operating systems. Therefore, created program is designed only for the GNU/Linux operating system.

Most important features of the created system are safety and reliability. The mail servers and the e-mails were never designed to store files [13]. In addition, e-mail security is dependent on the service provider. Our system avoids these limitations.

Adam Domanski
Institute of Informatics
Silesian Technical University
Akademicka 16, 44–100 Gliwice, Poland
e-mail: adamd@polsl.pl

Joanna Domanska
Academy of Business
Cieplaka 1c, 41–300 Dabrowa Gornicza, Poland
e-mail: joanna@iitis.gliwice.pl

A. Kapczynski et al. (Eds.): Internet - Technical Develop. & Appli. 2, AISC 118, pp. 201–208.
springerlink.com

This implementation was a form of experiment. The first goal was to create a working system that allows normal operation and files storage. The second objective of this study was to determine the possibility of practical use of this technology in the future.

18.2 The Linux Files System

In the operating systems based on the Linux kernel the file system mechanism was created in a very general way [1]. The data exchange between the process and the storage executes in the kernel. The kernel determines whether the process has access to files. This solution releases the processes of handling the file system and introduces the transparency. The process is not able to determine which file system provides the data.

To simplify the creation of the file system programs, the creators of the kernel introduced the VFS (Virtual File System) mechanism [8]. VFS is a kernel interface which provides a relatively small set of the system calls (e.g. open, close, read, write, mkdir, link). The implementation of the file system requires writing only a few specified above system calls. From the application point of view a place to download the data is not important. The application can read the data from the magnetic media, can download them through the network or from another location. Figure 18.1 shows the diagram of the VFS mechanism.

The support of the file system is localized in the system kernel. The process runs in kernel space and has unrestricted access to computer resources. In the disk file systems the program must have access to a physical device and only the kernel has such rights. The kernel program works without any limitations, with the highest priority and speed.

These aspects are not important for the network file system. Connecting to another computer using the TCP protocol does not require any additional rights. This can be done by each user process in the system. Speed is also not significant in this case. Delay associated with the data transmission over the network typically exceeds the delay associated with the lower priority process. This fact was used in the FUSE mechanism (Filesystem in USErspace). The FUSE is the interface between the VFS and the operating systems. The FUSE operates in the user memory space [7] and gives the possibility to implement and use the file system without influencing the kernel. This is important for the system security. Processes launched by a non-administrator user can not threaten the security of the system. The FUSE also provides the ability to mount a file system for ordinary users. Figure 18.2 shows the path of a filesystem. The FUSE kernel module and the FUSE library communicate via a special file descriptor which is obtained by opening the file: /dev/fuse. This file can be opened multiple times and the obtained file descriptor is passed to the mount syscall - to match up the descriptor with the mounted filesystem. The FUSE does not require to write software in programming language compiled into machine code. The software described in this paper was developed using Python language.

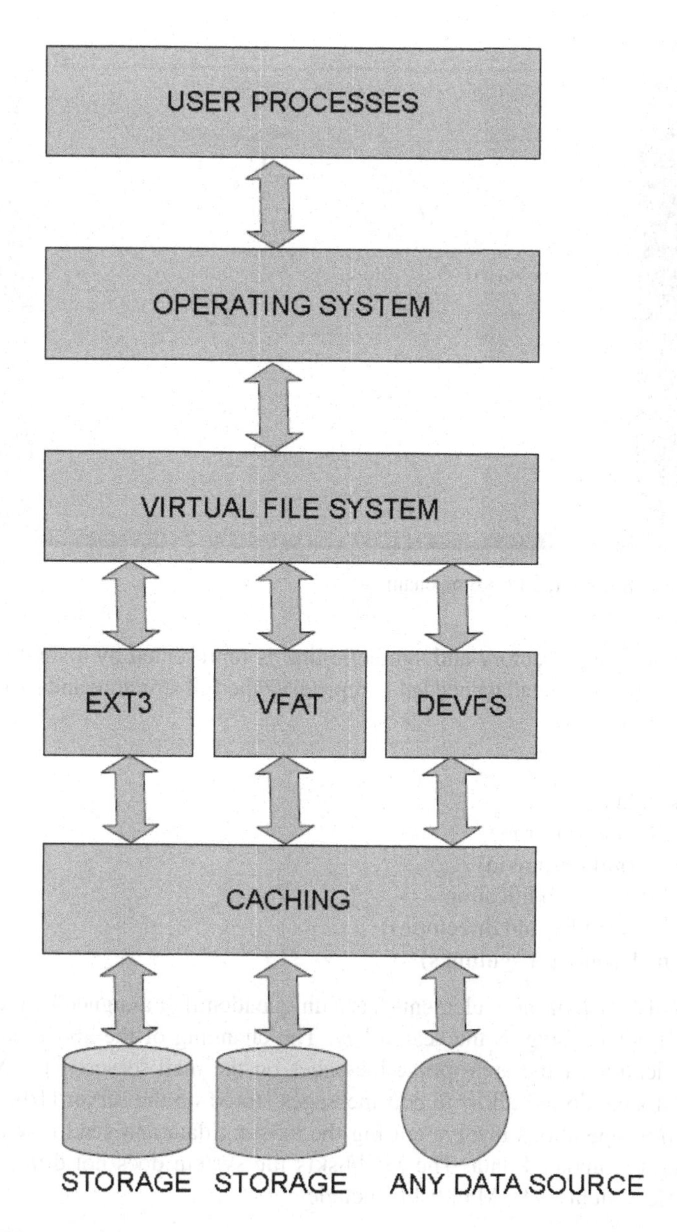

Fig. 18.1 Diagram of the Virtual File System mechanism

18.3 The File System Maildiskfs

The basic function of each file system is keeping data and structure of files and directories. In the case of our system all this information can be found on mail

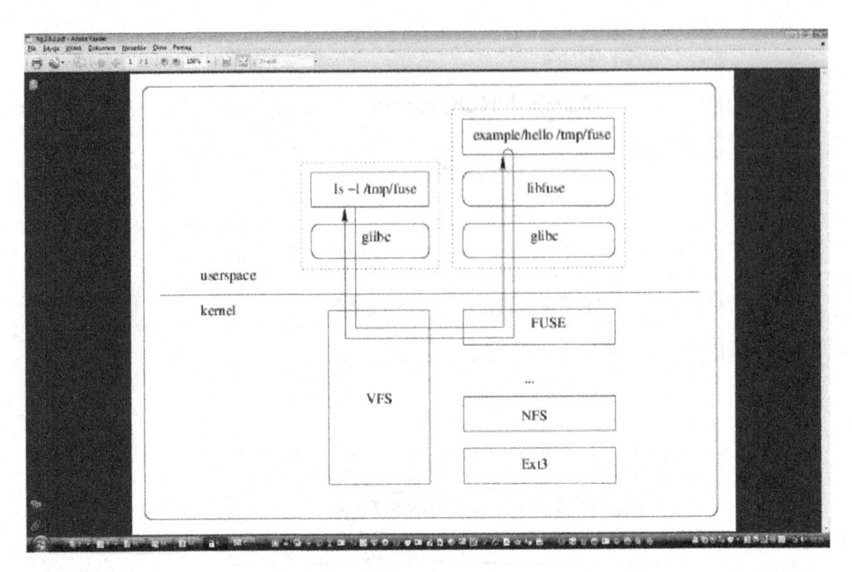

Fig. 18.2 Diagram of the FUSE mechanism

servers. Each file, directory and symbolic link is represented by a single message which contains the metadata needed to reproduce the full structure and system state. Metadata contains:

- path
- access rights
- owner ID (uid - user id)
- group ID (gid - group id)
- time of the last modification
- size (only for files and directories)
- target path (only for symlinks)

Additionally, each of these elements has a unique identifier assigned by the system. This identifier is using as the search key. The changing of the above data forces the modification of the appropriate messages on the mail server. The SMTP and POP3 protocols do not allow to edit messages stored on the server [10]. The data modification operations involve erasing the existing data and send new messages containing the updated data. The Maildiskfs file system does not differ from the popular file systems. The Maildiskfs includes:

- files of any size (their size does not exceed the capacity of the drive)
- arbitrarily nested directories
- symbolic links
- statistics of the System (space, number of existing files).

The program provides the full encryption of data and directory structure. The Maildiskfs encrypts the connection to the SMTP and POP3 servers. The system allows to create a single, continuous disk space on several mail servers. Program storages on the local computer only the configuration file. The configuration file is encrypted. In order to protect the system against the potential loss of the data one can back up the configuration file and store it elsewhere.

18.4 An E-mail Message

The message contains the information about the file consists only of the headers [13]. The content is added only to comply with the standard, but it does not contains any key information. The content type is specified in the "text/plain", or plain "text" [9]. Each message contains the following headers:

- X-Maildiskfs-Id - identifier of the file,
- X-Maildiskfs-Uid - ID of the owner of the item,
- X-Maildiskfs-Gid - the group ID element,
- X-Maildiskfs-Mode - the right of access to the file, written in decimal form,
- X-Maildiskfs-mtime - modification time Unix,
- X-Maildiskfs-Size - current file size in bytes,
- X-Maildiskfs-Path - the path to the file,
- X-Maildiskfs-MessageType - message type set to 'info'.

Message, which represents a fragment of the file, is a multipart message. One part of this message is a plain text, which - as in the previous case - does not play any significant role. The second part is the attachment marked as "application/octet-stream" (stream of octets) type [9]. The set of headers for this type of message includes:

- X-Maildiskfs-Id - identifier of the file,
- X-Maildiskfs-PartId - identifier of the fragment,
- X-Maildiskfs-SequenceNumber - the sequence number (the part of file),
- X-Maildiskfs-MessageType - the message type set to 'content'.

In addition, all messages sent by the system include the following, common headers:

- From - the name and address of the sender,
- To - the name and address of the recipient,
- X-Maildiskfs-MessageRevision - the version of the message,
- X-Maildiskfs-Validate - a string, allowing verification of the validity of posts,
- User-Agent - software id that drafted and sent the message (Maildiskfs / version of the software).

All Maildiskfs system headers (starting with the X-Maildiskfs) are encrypted before sending to the server. Then they are encoded into a transport form (binary to text) [9].

When a message is read from the POP3 server, the program checks the header "X-Maildiskfs-Validate". If it does not appear in the message or if its content is not successfully passed the verification process, the message is considered as invalid. This mechanism protects against a situation in which the structure or contents of the file system would be modified by an attacker by sending an appropriately prepared message with headers and content [17].

18.5 Performance Tests

System performance tests are based on a comparison with other similar solutions. Performance of any mechanism using the network depends on its current state. In the case of the low bandwidth and high latency (for example due to heavy load) the quality of all services decrease. These factors make it difficult to choice the best solution.

The tests have been developed in the ideal environment. The computer was located in a local area network. We performed the following tests:

- send file - size of 1MB
- send file - size 10 MB
- send file - size 100 MB
- download file - size 1 MB
- download file - size 10 MB
- download file - size 100MB
- sequence of read / write file - size 100 B

During the tests, we compared three different file systems based on FUSE:

- Maildiskfs
- curlftpfs [18]- a file system using FTP (File Transfer Protocol) and curl library
- sshfs [19]- using an encrypted connection through the SSH [6] channel

Table 18.1 shows the results of the tests. We have obtained very good results for the Maildiskfs system for small file sizes. This result is caused by using the cache. For each file upload was made after the download operations, so it can be expected that all or part of the data was retrieved from the cache. The queueing of sending e-mails events cause that the measurement period is ended and the file is still sending. However, the cache is extremely effective for frequent write and read operations and for the relatively small files. In the case of large files the cache performance drops dramatically.

Table 18.1 File systems performance

	Maildiskfs	curlftpfs	sshfs
1 MB upload	1.45 s	1.40 s	2.69 s
10 MB upload	15.29 s	5.04 s	20.12 s
100 MB upload	173.76 s	92.48 s	140.33 s
1 MB download	0.26 s	0.74 s	2.44 s
10 MB download	4.03 s	5.75 s	10.64 s
100 MB download	1345.18 s	53.51 s	143.36 s
100 B upload/download	0.05 s	0.48 s	0.23 s

18.6 Maildiskfs and the Spam Problem

Spam - the unwanted correspondence sent via e-mail has become one of the major problem of the modern Internet technology [16]. The financial losses due to this phenomenon are estimated at tens of billions annually. The rivalry between people sending unwanted items (spammers) and people trying to protect the users email box still continues nowadays. There are many mechanisms for analyzing and filtering email messages. The description of the used methods is beyond the scope of this work. However, the existence of anti-spam filters may have a negative impact on the correctness of the system. The specificity of the system causes frequent send and receive messages from the mail server. Such action is not natural for the mail servers and can be considered as abuse. A anti-spam filter based on message content analysis may qualify our mails as an unsolicited message. This is due to several factors:

- message does not contain any content,
- unknown mail headers appear in the mesagess,
- message has only one attachment with an unknown content,
- all messages come from the same sender.

The capabilities to protect our mails against the fact of recognition as unwanted messages are limited by the configuration of mail servers. If the mail server provides such capabilities, the user (before installing the system) should configure his mailbox to unconditionally accepting the message from the defined sender. For this purpose, the mechanism of white lists is used. In the case of e-mail the white list contains the addresses or entire domains from which the messages should be delivered without the spam filters. The number of security mechanisms are growing up. It is possible that the new mechanism cause wrong operation of the Maildiskfs system. For example:

- greylisting [14] - the mail rejection identified by three elements: the sender IP address, the sender e-mail address, the recipient e-mail address. This mechanism protects against transmission of messages by programs which are not mail servers.
- Sender Policy Framework [15] - the assignment an email domain names to specific IP addresses. Only the assigned addresses are authorized to send messages from the domain. This mechanism protects against using the direct messaging system.

18.7 Conclusions

This article shows the new storage data mechanism. The Maildiskfs system performance turn out to be surprisingly high. However, the system can be regarded as a threat to the messaging mechanism. Currently, the Internet world seeks to standardize and strict adherence to established rules. From this point of view Maildiskfs breaks many rules. Its action breaks the created order. However, the authors did not want to create a malicious program, but only tried to obtain a completely new functionality.

Presented and used technology in its current form seems to work quite properly. However, in the future this may change. The biggest impact on the Maildiskfs work may have anti-spam mechanisms, which are constantly developed and improved. We expect that in the near future, the effective anti-spam mechanisms will be created. If it happens the Maildiskfs system stops working or appears quite useless.

References

1. von Hagen, W.: Systemy plików w Linuksie. Wydawnictwo Helion (2002)
2. Praca zbiorowa: XML, Wikipedia, http://pl.wikipedia.org/wiki/XML
3. Montowanie, Wikipedia, http://pl.wikipedia.org/wiki/Montowanie
4. Cache, Wikipedia, http://pl.wikipedia.org/wiki/Cache
5. Full Disc Encryption. Wikipedia, http://en.wikipedia.org/wiki/Fulldiscencryption
6. Praca zbiorowa, Secure Sockets Layer. Wikipedia, the free encyclopedia, http://en.wikipedia.org/wiki/SecureSocketsLayer
7. Praca zbiorowa, FUSE Wiki, http://fuse.sourceforge.net/wiki/
8. Praca zbiorowa, Wirtualny System Plików, Wikipedia, wolna encyklopedia, http://pl.wikipedia.org/wiki/VFS
9. Praca zbiorowa, RFC 1521 - MIME (Multipurpose Internet Mail Extensions), http://www.faqs.org/rfcs/rfc1521.html
10. Praca zbiorowa, RFC 1939 - Post Office Protocol - Version 3, http://www.faqs.org/rfcs/rfc1939.html
11. anthony@interlink.com.au, pydns, http://pydns.cvs.sourceforge.net/pydns/pydns/README.txt?view=markup
12. Langfeldt N.: DNS HOWTO http://www.ibiblio.org/pub/Linux/docs/HOWTO/other-formats/pdf/DNS-HOWTO.pdf
13. Rafa, J.: Sztuczki z poczta, cz.2, (1999), http://www.wsp.krakow.pl/papers/procmail.html
14. Harris, E.: Nastepny krok w wojnie ze spamem: Greylisting (2003), http://projects.puremagic.com/greylisting/whitepaper.html
15. Sender Policy Framework Introduction, http://www.openspf.org/Introduction
16. Kucharski, P.: Spoleczne i prawne aspekty spamu, http://42.pl/esej/esej-o-spamie.pdf
17. Collins-Sussman, B., Fitzpatrick, P.C.M.: Version Control with Subversion (2002-2007), http://svnbook.red-bean.com/en/1.4/svn-book.pdf
18. CurlFtpFS - A FTP filesystem based on cURL and FUSE, http://curlftpfs.sourceforge.net/
19. SSH Filesystem, http://fuse.sourceforge.net/sshfs.html

Chapter 19
Threats to Wireless Technologies and Mobile Devices and Company Network Safety

Teresa Mendyk-Krajewska, Zygmunt Mazur, and Hanna Mazur

Abstract. The wireless technologies have become so attractive that more and more networks, including company ones, are created according to these standards. Mobile access to network resources is also becoming more and more common. Unfortunately, security mechanisms which are used nowadays show a number of faults and are prone to attacks, which, in the age of growing threats, including threats to mobile devices, raises concern about the future of wireless realizations. It is particularly significant in the case of industrial networks due to their frequent strategic importance. Management personnel should be aware of real threats whereas employees should strictly obey company safety policy principles assumed. The aim of this article is to show the importance of the problem and a scale of this phenomenon.

19.1 Introduction

The attraction of wireless realizations and mobile connection to the Internet or local company networks are commonly known. Available mobile devices, which are more and more functionally improved and, at the same time, cheaper and cheaper, are raising justified interest among potential users. They are eagerly used for data storing, banking services or business task realization and enable remote access to company resources. With their rising popularity, they are becoming a more and more attractive target for attacks of internet criminals. Although, with the development of the mobile device functionality, a dynamic increase in threats to those devices is observed, many users are not aware of that or they do not attach due weight to this fact. At the same time, more and more mobile devices have access to Wi-Fi (company, domestic) networks, which increases a possible infection of these networks.

Teresa Mendyk-Krajewska · Zygmunt Mazur · Hanna Mazur
Wroclaw University of Technology, Institute of Informatics
Wyb. Wyspiańskiego 27, 50-370 Wroclaw, Poland
e-mail: {teresa.mendyk-krajewska,zygmunt.mazur}@pwr.wroc.pl,
hanna.mazur@pwr.wroc.pl

A. Kapczynski et al. (Eds.): Internet - Technical Develop. & Appli. 2, AISC 118, pp. 209–225.
springerlink.com

Experts warn holders of such devices as smartphones or smartbooks that they will soon experience the same problems as those associated with computer usage. Thus, the use of wireless technologies and easy access to the Internet from mobile devices pose a number of additional threats to their safe operation. The attacks undertaken aim, for example, at disturbing operation or blocking the usage of a given network or device or obtaining unauthorized access to their resources.

19.2 Threats to Wireless Technologies

For the realization of wireless connections, the Wi-Fi technology may be applied (a set of standards for building computer networks) as well as the Bluetooth technology (for data transmission over short distances between electronic devices such as phones, computers, palmtops, etc.), the GSM technology (*Global System for Mobile Communications*) used in cellular networks, or a technology that creates a separated network other than those defined for the above-mentioned standards.

Signal transmitted in an open space is more prone to threats than the one transmitted through cable due to easier access to a transmission medium (it may be disturbed or received by any station which is within range of a transmitting station). Despite the application of various security mechanisms such as signal range limit, hiding of network identification, MAC (*Media Access Control*) address-based access control and the use of available cryptographic solutions in order to ensure data confidentiality and integrity as well as authentication of communicating parties – wireless networks are not safe. Improper installation and configuration of a network may already be the cause of problems. The basic errors of this type are, first of all, the following:

1. failing to hide SSID/ESSID network identifier (Service Set Identifier/Extended SSID),
2. too simple name of a network is used,
3. lack of control over a connection area (range may be too large when the power of signal or the direction of electromagnetic wave spreading are improperly adjusted),
4. access points are left without passwords,
5. failing to apply the strongest available safeguards.

Among the actions which are harmful to wireless networks the following may be listed:

1. signal jamming/disturbing through emission of waves of the same or similar frequency and relevant power,
2. installation of false access points,
3. overloading a system through sending in a large number of packets to be served,
4. attacks on cryptographic safeguards,
5. disrupting the client-access point connection (which is considered preparatory for further attacks),
6. Man-in-the-Middle attack.

The Wi-Fi wireless networks are easy to be detected and analyzed, with an additional threat posed by an availability of ready-made analytical tools, which enable the following:

1. analysis of packet headers and initialization vector fields,
2. peeping at network names,
3. checking MAC addresses,
4. identification of cryptographic safeguards used or pointing out the lack of them,
5. signal range determination,
6. obtaining information on IP (*Internet Protocol*) addresses.

Less or more complex tools are intended for network testing – from simple ones which perform a relevant function, to complex systems which support safety management (unfortunately, they can also be used to carry out an attack). Testing covers the following but is not limited to it: correctness of access point configuration, control of the authorized user list, signal range checking, the presence of illegal access points, or the existence of other nearby networks. Among the sniffing tools (sniffers) intended for wireless networks, the following may be specified: Mognet, snort, tcpdump, airfart, AiroPeek, Ettercap, dSniff, Wireshark (earlier Etheral), nmap, and NetWitness.

For the safety of data transmission, various cryptographic safeguards are applied. For the Wi-Fi networks, these are WEP (*Wired Equivalent Privacy*), WPA (*Wi-Fi Protected Access*), and WPA2 standards. It is only the WPA2 which provides sufficiently strong safety mechanisms, however, it is not compatible with earlier solutions. WPA is prone to attacks because the TKIP protocol (*Temporal Key Integrity Protocol*) is cryptographically weak, authentication is one-sided, and the RADIUS protocol faults (*Remote Authentication Dial-In User Service*) enable skipping the authentication process. As far as the WEP is concerned, it should not be applied at all now due to its serious faults (proneness to attacks on the immutability of the key and to attacks with the use of the known initialization vector). The Wi-Fi Alliance announces that it will soon withdraw the WEP and TKIP protocols [1] from the new equipment. Known weaknesses of the cryptographic protocols and algorithms have been used to work out tools which enable carrying out the attack. The popular tools are as follows: AirSnort, which enables recovery of the WEP key on the basis of an analysis of the sufficient number of the intercepted packets (e.g. by means of the FMS method – which name derives from the surnames of its authors: Fluhrer, Mantin, and Shamir), Aircrack (which uses a dictionary attack on WPA but is not limited to it), and WepLab, which uses optimized FMS and Korek's attacks. The Wigle.net site has worked out the list of 1000 most popular WLAN network names in the world, and it was used by the makers of tools for cracking safeguards as the base for creation of special tables (*rainbow tables*), which facilitated the practice. Each of these networks has access to a table that contains initially calculated values facilitating detection of the cipher key for the WPA and WPA2. The table can be downloaded from different sources.

Apart from the attacks on cryptographic systems, there are also attempts to carry out a DoS attack (*Denial of Service*), and blocking access to a company network may have serious consequences. In the case of the WLAN network (*Wireless Local Area Network*), the DoS and DDoS (*Distributed DoS*) attacks may regard fake authentication packets – the *Authentication DoS Mode* attack or switching over a network card to a continuous transmission mode – the *Queensland DoS* attack).

On the other hand, implementations of protocols and a lack of the base station authentication are the weakest elements of the GSM system (which covers around 80% of the mobile telephony global market). The GSM protocols are prone to the *Man-in-the-Middle* attack with the use of the fake BTS base station (*Base Transceiver Station*). In the cellular networks, the A5 stream cipher is used, which versions A5/1 and A5/2 are already unsafe. In 2008, experts worked out the method for cracking the A5/1 cipher, and now, with the use of commonly available cheap devices and software, it can be done for the time shorter than half an hour, and the range of tapping amounts to 20 km. The methods for cracking the A5/1 and A5/2 ciphers were already elaborated in 1998, however, their high costs constituted a barrier for their practical application. In metricconverterProductID2007, a2007, a stronger version of the cipher – A5/3 was published. Unfortunately, the cipher was cracked already in 2010 [2]. Faults in the mobile device software and network protocol packets which are inconsistent with specifications are great threats. Not too much attention is drawn to the gaps in the GSM transmission protocols, which may be surprising considering the massive popularization of this technology. All safety risks for the GSM network users are not publicized. The Polish operators try to introduce new wireless technologies successively, for example UMTS (*Universal Mobile Telecommunications System*), which is considered to be a safer solution. Networks which are realized within these standards may cooperate with each other.

Attacks on the Bluetooth technology make use of faults in protocol implementation and faults in the phone and controls software. For carrying out those attacks, special hacking tools have been worked out. The Bluetooth technology is considered to be unsafe so it is not recommended to be used for sending confidential information. Thanks to it a device is visible and open for data exchange with other users within up to a twenty-metre radius.

The usage of wireless technologies is associated with a lot of threats therefore due to their dynamic development and the need for achieving a high level of security, they require new safety standards to be worked out.

19.3 Functionality of Placemobile Devices

While the Microsoft Windows operating system has a dominant position in the personal computer market, in the case of mobile devices, there is a lack of such an outright leader. Devices with Windows Mobile and Windows Phone 7 systems (manufactured by Microsoft), iPhones (Apple) with the iOS operating system are very popular as well as devices with the Symbian system (Nokia), the Brew system (Qualcomm), and phones with the S/E GOLD (Siemens) or Linux Mobile platform. On the other hand, in the American market, smartphones with the Android platform

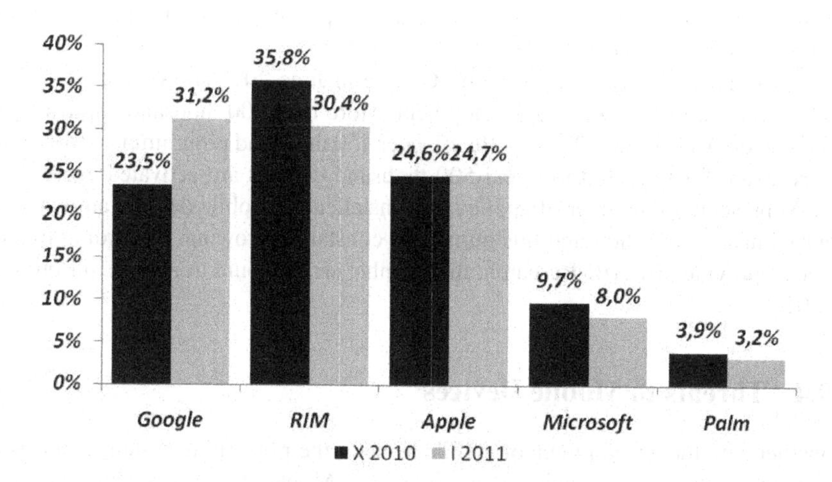

Fig. 19.1 The most popular smartphone system platform providers in October 2010 and January 2011

manufactured by Google and RIM's (*Research In Motion*) devices operating under control of BlackBerry OS are the main leaders.

Percentage distribution of the most popular smartphone system software providers in the USA at the turn of 2010/2011 is shown in Figure 18.1 [3].

Functionality of cellular phones is systematically developed. Apart from the use of the basic functions such as voice communication, text communication (SMS, *Short Message Service*), sending images (MMS, *Multimedia Message Service*), and data storage, users have additional options at their disposal, for example, electronic mail, Internet communicators, calendar with a planning function and a function of browsing tasks from other devices, calculator, digital camera (with flash), film recording, direct access to social networking websites, remote data deletion, or the GPS navigation system (*Global Positioning System*).

In spite of a wide range of operating systems for mobile devices, the work on new solutions is carried out continuously. The Chinese company Baidu, which is working on its own operating system, is a good example. On the other hand, Microsoft introduced a newer version of the Microsoft Windows system called Windows Phone 7 in 2010 and resigned from further development of the Windows Mobile system, which has not been updated for a long time (which does not mean a drop in the interest of virus creators in this system).

On the other hand, Apple announced its plans to introduce an iOS 5 mobile system in autumn at the conference in placeCitySan Francisco in June 2011. The system will have a lot of new solutions, including an iCloud service (a possibility to store multimedia data in *cloud computing*) and a possibility to use the iTunes Match service, which task is to search for music files illegally installed on users' discs and to add them to the offer of the Apple on-line store. Unfortunately, in this case, the whole contents of the user's device will be scanned simultaneously. Apple also works on blocking iPhones in order to make it impossible to record films and take photos during cultural events [4].

For example, the data published by Google in June 2011 shows a scale of the mobile application and device development. More than 200 thousand applications available on Android are already offered, over 450 thousand programmers work out the software for this platform, and 500 thousand devices are activated every day [5]. At present, the number of applications installed on mobile devices amounts to approximately 16 billion and this number is constantly growing. According to the forecast provided by ABI Research, this number will amount to approx. 44 billion in 2016.

19.4 Threats to Mobile Devices

Together with the development of mobile devices, their operating systems, and user applications, the threat to their safe use increases. Mobile devices are objects which are easy to be stolen by the outsiders due to their small size, storing within easy reach, and their use in public places.

Harmful software was initially directed to a specific system platform. The first mobile phone virus was detected in June 2004, and, already at the end of the same year, the estimated number of harmful programs amounted to one hundred. The problem of mobile device infection has become significant only after working out cross-platform viruses.

An (IncognitoRAT) worm written in Java script which attacks both MAC and Windows system has been detected recently. This software when properly prepared may also attack iPhones and iPads [6].

The Java 2 Micro Edition (J2ME) system was a smartphone mobile system which was the most threatened with attacks in 2010 – 57.67% attacks, which is proved by the results of the research 'Evolution of the mobile harmful software: metricconverterProductID2010Š2010' carried out by the company Kaspersky Lab [7]. The percentage specification of attacks on the smartphone operating systems in 2010 has been shown in Figure 18.2.

The present popularity of Android has caused an immediate growth of the harmful software which attacks just this system. The company Lookout Mobile Security has recently identified the jSMSHider Trojan horse virus intended for the Android system with great potential for action, which enables remote message management and application installation [8] but is not limited to it.

Harmful mobile phone software may perform a lot of illegal actions such as: blocking memory or device cards, file infecting, installation of harmful software, data modification or deletion, data theft (ex-directory phone numbers, company documents, bank account access passwords), enabling remote access to a device (and thereby e.g. to a corporate network or e-mail), sending SMS or MMS messages, downloading files from the Internet, switching off system safety mechanisms, including antivirus programs.

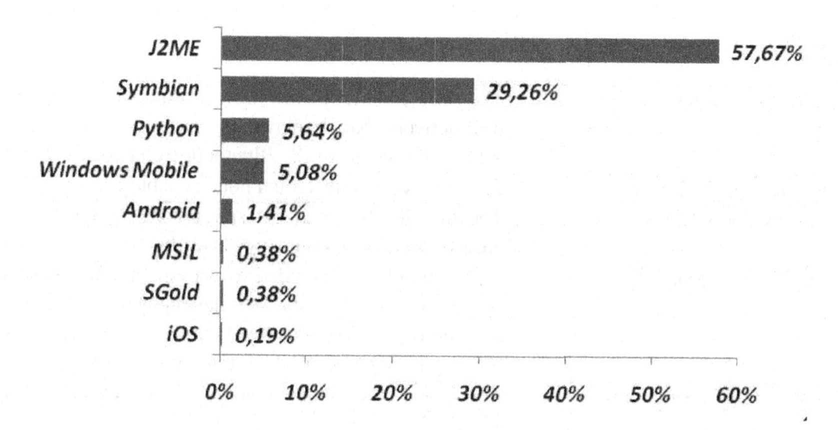

Fig. 19.2 Percentage specification of the frequency of attackson smartphone operating systems in 2010

Among the first Trojan horse viruses, the following may be listed: Locknut.C (which pretends to be a patch for the Symbian system phone software), Skuller (blocks a device with the Symbian operating system and enables only carrying on phone talks), Skulls.S (places a few copies of Cabir.F virus on a device) as well as Cardtrap.F and Cardtrap.G (deactivate Symbian system applications and install malware that attacks the Windows system) [9].

Examples of popular mobile phone threats have been included in Table 18.1.

A device can be infected with a harmful code in many different ways, for example, during downloading files from uncertain Internet sources (e.g. jingles, graphics, music files, games or applications), while sending SMS and MMS messages, or through the Bluetooth technology. There are a lot of types of attacks which use the Bluetooth technology, e.g. BlueBug – which allows taking control over a device (through the use of hidden RFCOOM channels), Blueprinting – which consists in collecting information on active devices, or BlueChop – which leads to disturb operation of other devices located within a given subnetwork. It is also possible to download a virus on a memory card during synchronization of data with a stationary computer; this is how e.g. Cardtrop and Mobler viruses operate. A system may also be infected while downloading applications from a legal source that already contains a harmful code. In May 2011, infected programs (as a newer version of the application) were placed on Android Market website, which after their installation, conveyed information on a given phone to an external server, enabling taking control over the device [10].

Unfortunately, harmful codes and the actions taken up by them are often camouflaged, and it is performed in a more and more refined way.

Not all devices are equally easy target. For example, in order to infect an iPhone (to install relevant applications), one should break safeguards first. Gaps in the software safeguards are the main threat to this platform, which enable unauthorized access to the devices. At the beginning of 2009, a gap in a few Symbian versions

Table 19.1 Selected threats to mobile devices

Cabir (2004, Symbian)	The first mobile phone worm which does not act destructively but displays the message 'Caribe'; it spreads by means of the Bluetooth technology
Pbstealer (2005, Symbian)	The first cell phone Trojan horse; enables data theft
BBproxy(2006, BlackBerry)	The first RIM (backdoor) smartphone Trojan horse, enables attacks on company networks
RedBrowser(2006, J2ME)	The Trojan horse intended for various mobile devices (smartphones, palmtops, cellular phones); offers free sending of an SMS message yet what it really does is sending it to premium rate numbers
Comwar (2007, Symbian)	The first mobile phone virus; it spreads by means of MMS
InfoJack (2008, Windows Mobile)	The Trojan horse that transmits back device data; switches off warning against starting and installing uncertified programs
RickRoll(2009, Apple)	The first undangerous iPhone worm (displays a photo of Rick Astley, a singer) which cracks safeguards in iPhones but does not change SSH (Secure SHell) password; based on it, the Ike worm was created to create a botnet and steal authentication data for online banks
Pornidal (2009, Symbian)	(Dialer) program for dialling out international exdirectory phone numbers
Yxe or Sexy View(2009, Symbian)	A worm spread by means of SMS messages with links to erotic websites, undersigned with a legal certificate
Terdial (2010, Windows Mobile)	The Trojan horse that dials out international numbers spread together with 3D Anti-Terrorist game
Stuxnet(2010, Windows)	Very complicated; the first worm which attacks SCADA industrial networks (which control production lines, pipelines, power plants but are not limited to them) conveyed by means of pendrives
FakePlayer(2010, Android)	The first Android systemTrojan horse, installed together with multimedia files in order to send costly SMS messages
Gemini(2010, Android)	Harmful software (malware) offered as a safe application
DroidDream(2011, Android)	A virus that has infected applications at Android Market
ZeuS(2011, Symbian, BlackBerry)	The Trojan horse (stealing data for online bank accounts) known before from its attacks on computers; it proposes to install a program that increases a processor operation, but instead it installs a harmful software that redirects one to fake bank websites.

was detected, used by exploit '*The Curse of Silence*''. If a specially prepared SMS message (invisible to a user) is sent to a device with such a system, it will not be possible for the device to send and receive SMS and MMS messages – but apart from that, it will function in the ordinary way, which delays detection of the irregularity. A gap in the mobile Android VoIP (*Voice over Internet Protocol*) application is another example, which enables stealing e-mail addresses, contact list, and chat history of Skype users.

On the other hand, a different gap in Google's Android platform causes such a problem that a fabricated SMS may lead to the loss of connection with the network, as a result of which it is possible to carry out a DoS attack. Examples of viruses which use different system gaps are as follows: Trojan.SymbOS.Skuller, Trojan.SymbOS.Dampig, Trojan.SymbOS.Romride, and Net-Worm.IphoneOS.Ike.b.

The so-called SMS phishing poses a big threat to mobile devices. In this case, anonymous senders send SMS messages in which users are asked, under the pretext, for making a phone call to an indicated phone number. Next, during the realization of a given connection, they are asked about confidential information. The other examples of a socio-technical attack are different ways in which money is wangled out of the users. A message sent under the pretext of notifying the user about a reward to be granted to him/her after sending a (costly) SMS message to a given number may also be an abuse. Other phenomena observed are as follows: persuading the users to download paid files from the Internet, encouraging the users to take part in various contests, enforcing users to make a costly phone call, SMS message automatic sending or initializing connections. Sending messages to costly phone numbers without the user's knowledge constitutes about 35% of all threats identified by Kaspersky Lab.

The increasing problem of threats has been noticed by cellular network operators, who are already informing their subscribers about a need to exercise caution in case when a ringing party hangs up after one ring, or when received SMS or MMS messages come from an unknown sender. The greater functionality of a mobile device, the bigger threats to safety and more serious consequences as a result of its loss or virus infection there are.

Observed threats to the mobile devices are often a type of local attacks which occur in specific countries or regions. In 2009, Russia, Indonesia, and the West European countries belonged to those with the biggest threat, whereas in 2011, Sudan, Bangladesh, and placecountry-regionIraq are listed among them [11, 12].

19.5 Threats to Mobile Devices in Practice

Nowadays, smartphones are developed devices which are more like a computer than a telephone. They are intended to store data, perform online financial transactions, do online shopping or use electronic mail. Hence, network criminals are interested in them and the number of harmful programs intended for those devices increases.

Trojan-SMS.WinCE.Sejweek may serve as an example of a Windows Mobile platform Trojan horse, which has operated since 2009 and which, after successful connection to a server, downloads an XML file with the following model contents to a device [7]:

```
$<$?xml versión=''metricconverterProductID1.0''1.0''
encoding=''Windows-metricconverterProductID1251''
1251''?$>$
$<$getxml$>$
$<$phone$>$YGLYGLMKTYGL$<$/phone$>$
$<$text$>$text$<$text$>$
$<$interval$>$IUNIUNYGLYGLIUNIUN$<$/interval$>$
```

Ciphertexts which follow the tags <phone> and <interval> are deciphered according to a definition stored in the code of Trojan horse:

```
public static void FillCodeTable()
{
codeTable = new GeneralIDS.ShifrDataTable();
codeTable.AddShifrRow(''YGL'',
''metricconverterProductID1''1'');
codeTable.AddShifrRow(''HKR'',
''metricconverterProductID2''2'');
codeTable.AddShifrRow(''DPO'',
''metricconverterProductID3''3'');
codeTable.AddShifrRow(''WHR'',
''metricconverterProductID4''4'');
codeTable.AddShifrRow(''MKT'',
''metricconverterProductID5''5'');
codeTable.AddShifrRow(''PQA'',
''metricconverterProductID6''6'');
codeTable.AddShifrRow(''LOO'',
''metricconverterProductID7''7'');
codeTable.AddShifrRow(''THU'',
''metricconverterProductID8''8'');
codeTable.AddShifrRow(''XRE'',
''metricconverterProductID9''9'');
codeTable.AddShifrRow(''IUN'',
''metricconverterProductID0''0'');
}
```

In this case, it means sending an SMS to the phone number 1151 every 11 minutes (without the phone owner's knowledge). In 2009, it was the ciphered phone number 7122, an SMS sent to it cost 10$. A parameter following the tag <interval> determines time in the form of hhmmss.

These types of Trojan horses are installed on a phone e.g. during downloading a file with a game, music, or a photo. In 2010, the Trojan horse.WinCE.Terdial was downloaded together with the free game 3D Anti-Terrorist (intended for the Windows Mobile system), and dialled out six international numbers every month (including calls to Somalia and the Dominican Republic) thanks to the automatic starting function called CeRunAPpAtTime. The first worm (Net-Worm.IphoneOS.Ike.a) for an iPhone with cracked safeguards and the unchanged default SSH password identified in November 2009, stole the owner's data, enabled taking control over a device, searched for new IP addresses and installed itself on those devices which were prone to infection. Operation of this worm allows building up a mobile network of botnets, which may be used by the criminals who control it to perform unpunished malicious actions [13]. A symptom that indicates the device infection is quick running down of a battery (because, if there are many wireless networks within a range of the device, the worm tries to send out its copies, which uses up the battery quickly). In November 2010, people who cracked safeguards without alternation of the default password and downloaded the worm Net-Worm.IphoneOS.Ike.b, and, then, logged into the Dutch ING Direct Bank, fell victim to phishing. As a result of that, they were redirected to a fake bank website where they unconsciously disclosed their logins and passwords [7].

In July 2011, the news about hacking into some mobile phones has been announced. The information read from SMS messages and the voice mail was used, then, in the press publications (in the daily paper "News of the World").

In May 2011, an author's survey regarding safety of smartphones was carried out among the students of Wroçşaw University of Technology. The research covered the group of 144 people. As many as 72% of the students surveyed experienced incorrect operation of the system, almost one third experienced a loss of data, and 17% confirmed the presence of viruses. In spite of that, only 16% of the smartphone owners use security software. Complete results regarding unfavourable phenomena observed are shown in Figure 18.3.

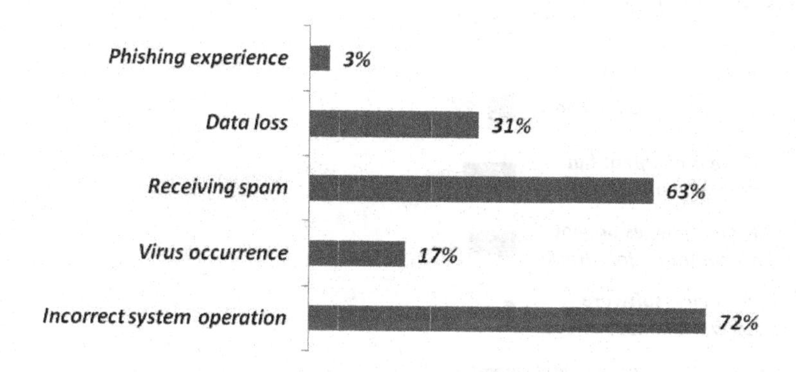

Fig. 19.3 Undesirable events observed by survey respondents

A loss of a device was experienced by 25% of the survey respondents. The percentage distribution of using the device to store or transfer confidential information by the surveyed users has been presented in Figure 18.4.

On the other hand, the research carried out by the company F-Secure shows that 75% of the smartphone users are afraid of losing data (e.g. contact addresses,

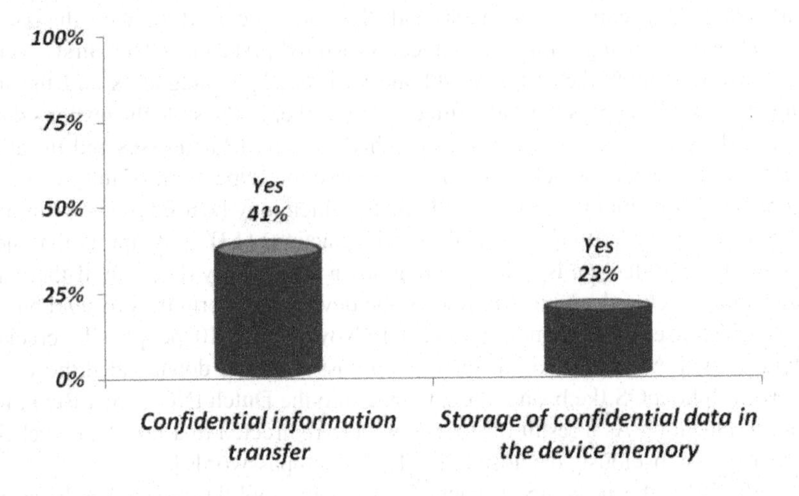

Fig. 19.4 The use of mobile devices to store/transfer confidential data

photos), whereas the Polish people underestimate this threat (only 15% say they are aware of threats) [10]. Meanwhile, it has been estimated that around 70 thousand new viruses, Trojan horses and other types of malicious mobile phone applications were created every day in the first quarter of 2011 [14].

However, the Polish users are aware that modern phones are the ideal target for virus makers (68% positive answers), which results from a survey carried out on-line within a group of more than 6 thousand people [15]. Its complete results have been presented in Figure 18.5.

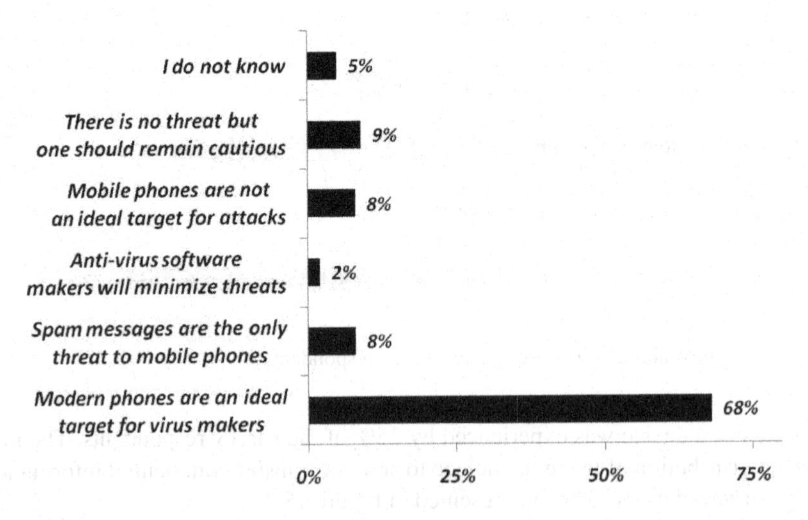

Fig. 19.5 Results of the poll regarding awareness of threats to mobile devices

19.6 Mobile Accessibility of Network Resources

Wireless connections improve comfort at work but, first of all, enable communication and data sending in an inaccessible area, where realization of cable networks would not be possible (when distances between objects are too long or an area is considerably diversified). In addition, such solutions are characterized by comfort and low installation costs, hence an interest in realization of a network with the use of the wireless technology is increasing. A possibility to access specific company networks, thanks to mobile devices, from any place outside a seat of a given organization, increases both work comfort and employees' activity, which makes their work more effective. For these reasons, wireless technologies offer great prospects, and mobile devices raise a tremendous interest. Unfortunately, the problem of safety threats, known with regards to the usage of conventional teleinformation systems, refers also, but not to a lesser extent, to the mobile equipment. Therefore, additional problems result from remote accessibility to company network resources.

The so-called. wardriving is a dangerous phenomenon for wireless WLAN networks, which consists in collecting information on these networks with the use of a car, a laptop, and a GPS receiver. For this purpose, high functionality tools are used. Sample wireless networks scanners which identify access points are the following: NetStumbler, MiniStumbler, Kismet, WifiScanner, and Gtkskan.

The problem of network safety threats acquires particular significance with regards to industrial networks and programmable electronic systems. The industrial networks can be realized as a whole, with the use of wireless technologies, or cable solutions may be combined with the WLAN networks used in a company. Elements of teleinformation networks are more and more often used in the industrial networks. In addition, both types of networks can be combined within the premises of a given enterprise, first of all, because of a need for direct accessibility of information for other systems, for example, the management support systems. That is why it is much more difficult nowadays to ensure the required level of safety for the industrial systems. Easier and easier access to these systems is becoming also a problem, among other things, due to the use of communication channels commonly applied for the purposes of network control and management.

There are errors in the Windows operating system which enable unauthorized persons to switch off remotely the SCADA system (*Supervisory Control And Data Acquisition*), which supervises the run of technological or production processes. Gaps in the industrial software are rarely reported, which does not mean that it is entirely free from any defects. In March 2011, the information on 24 gaps in the SCADA systems, provided by different manufacturers, was published. It should be highlighted that a lot of these systems were installed a long time ago and they have not been updated until today (in practice, it is not always possible to be executed) [12].

It is also more and more often allowed to have remote access to the network resources from the mobile equipment (from both company and private devices).

The use of wireless technologies offers great potential, hence a lot of equipment is designed just for industrial applications due to its suitable dimensions, functionality or ergonomics. Motorola or CSI (*Computer Systems for Industry*) mobile terminals may serve as good examples. Such devices are more and more frequently used by medical services, police or fire service. In some devices (industrial panels), the Haptic Touch technology (introduced by the company Saia-Burgess) is applied, in which a pressed element produces relevant vibrations. Such types of devices are intended for people who work in gloves, people whose work involves movement (the army), and those who work in the dark or are partially sighted.

Company networks to which wireless remote access is permitted, require particularly strong safeguards. Already in 2006, Nokia introduced a Sourcefire device enabling protection of company networks against outside threats in connection with the use of mobile equipment such as laptops or smartphones. The following three components: *Sourcefire Intrusion Sensor for Nokia* (for incoming traffic inspection and anomaly detection), *Sourcefire Real-Time Network Awareness for Nokia* (for network monitoring) and *Sourcefire Defense Center for Nokia* (for identification of long-term trends) enabled effective blocking of threats. According to NSS Lab, an independent testing organization, the Sourcefire device is characterized by the highest effectiveness in network attack detecting and, since 2007, it has been rated by the company Gartner among IPS (*Intrusion Prevention System*) system leaders.

The report '*Mobility and Security. Dazzling Opportunities, Profound Challenges*' proves that 95% of companies have elaborated a policy of using mobile devices connected to the network and have made their employees familiar with it, yet, only one third of the employees conform to the elaborated recommendations. A case of losing a mobile device is confirmed by 40% of the companies. Recorded data has been often lost irrevocably as backup copies have not been created [16].

An example of a quick reaction to an attempt to attack network resources may be the decision of the U.S. Lockheed Martin arms concern taken in May 2011, which switched off a service that enabled remote access to the company information resources with the use of tokens developed by RSA Security. The reason for this procedure was the effective attack on RSA servers carried out in March 2011, which resulted in the theft of confidential information enabling weakening of the SecurID authorization system (the algorithms used for password creation have probably been disclosed) [17].

The infected computers are used to send out spam messages, to carry out DDoS attacks or to use their computing power to crack cryptographic keys or for other purposes. There are many opinions that as the mobile device technology develops, if the connection to the global network is faster and faster, the mobile devices will become the same target as computers for network criminals, as considerable material benefits lie behind this illegal practice.

19.7 Mobile Accessibility of Network Resources

Until recently, the attacks on company networks were carried out mainly by unloyal employees who knew both company systems and their safeguards. Nowadays, thanks to the developed and commonly available tools, attacks may be also conducted by people who do not have any connections with a company and who do not even need to possess sufficient knowledge about the company.

Accessibility of network resources from mobile devices broadens the possibility of action, therefore the number of people (including also criminals) who are interested in new technologies will grow. As it is very easy to carry out an attack on poorly safeguarded wireless networks and due to the fast development of threats to mobile equipment, the strongest protection mechanisms which are available for users ought to be applied, and the protection should concern all platforms: from the hardware platform to the application platform. In the case of Wi-Fi networks, it is recommended to apply WPA2 (with the AES cipher algorithm) with the use of CCMP (*Counter Mode with Cipher Block Chaining Message Authentication Code Protocol*) and authentication pursuant to standard 802.11x, however, the highest level of security is provided by appropriately long cipher keys.

At present, a lot of applications intended for smartphones and mobile phones are offered, which purpose is to ensure safety for these devices. Their main functionalities are the following: protection against malicious software (spyware, malware), the possibility to locate a device (in case of its loss), and the possibility to create copies of backup data stored in the device memory. Of course, in each case, there is a possibility of regular automatic updating of the software and enforcing its periodic scanning. An example of such a type of application is Lookout Mobile Security intended for telephones with the Android system installed (from version 1.5) or Kaspersky Mobile Security offered for different models of devices and systems (Symbian, Windows Mobile, Android, and BlackBerry) [18]. On the other hand, the F-Secure Mobile Security 7 packet enables remote blocking of a mobile phone, deletion of all data from the device memory, activating the ring even with a set mute option, and localization of a device. The newly-introduced functions aim at providing safety for the mobile device usage.

The loss of a device as a result of theft or losing are commonplace Therefore, for example, the Mobile Tracker technology is applied in Samsung mobile phones, which enables fast localization of the device. If the owner of a given phone gives a phone number to which a relevant message is to be sent, then, in case of a change of the SIM (*Subscriber Identity Module*) card, an SMS containing the IMEI (*International Mobile Equipment Identity*) identification number and the number of a new SIM card installed in the lost phone will be sent to him/her automatically and unnoticeably. The data enables identification of a current mobile phone user.

Experts have also worked out a tool (the Ontrack Eraser Mobile program) intended for popular Symbian and Windows Mobile system platforms, which allows remotely, through an SMS, deleting data from the device in case of its loss. The contents of the inner memory of the phone as well as information stored on the memory card and the SIM card are destroyed irrevocably.

According to the assurances of mobile device providers, if a user does not perform jailbreaking (i.e. delete legal safeguards) and does not install software coming from an uncertain source, then, thanks to the built-in safeguards, the device should not be infected with a malicious code. However, the experience concerning the computer system usage shows that gaps, which undoubtedly exist in the mobile device software used by hackers for breaking into phones, will always pose threat to the safety of mobile devices.

19.8 Conclusion

Information systems are used not only by companies which run business activities but also by hospitals, medical facilities, power plants or public utility companies. One should be aware that these systems are threatened with an additional possibility of interference in the network resources with the use of mobile devices. The attacks on company networks with the use of wireless technologies is particularly dangerous as an offender remains anonymous.

The greatest anxiety is raised by the forecasts of an increase of interest in industrial espionage and attacks carried out on industrial networks, which aim at stopping or destroying business and public utility facilities such as power plants, public transport, traffic lights systems, dams, and hospitals, will intensify. Stuxnet is a known malicious code that allows blocking not only the Internet networks but also poses such a serious threat to the industrial networks. An attack with the use of this software has already been carried out on the placecountry-regionIran industrial networks (including the computers of the nuclear power plant). The public was informed about the attack in September 2010. Serious damage was not done at the time, however, according to experts, a real danger exists that industrial networks will be attacked with the use of computer viruses, and wireless access to the network resources increases it even more.

In order to minimize the threats to the information systems, both international standards organizations and regional and national organizations elaborate norms, standards, and recommendations which constitute the basis for creation of appropriate safety policies and implementation of adequately strong security systems. Examples of such elaborations are the following documents: ISO/IEC 15408-1:2009 *Information technology – Security techniques – Evaluation criteria for IT security – Part 1: Introduction and general model*, ISO/IEC 27001 *Information technology – Security techniques – Code of practice for information security management*, NIST SP-800-82 *Industrial Control System Security*, NIST SP-800-53 *Recommended Security Controls for Federal Information Systems and Organizations*, NIST SP 800-57 *Recommendation for Key Management*.

The management personnel should be aware of all the threats and properly elaborate the company safety policy, whereas the employees should strictly obey the principles assumed.

References

1. Kingsley-Hughes, A.: Wi-Fi Alliance to dump WEP and TKIP, not soon enough (2010), `http://www.zdnet.com/blog/hardware/wi-fi-alliance-to-dump-wep-and-tkip-not-soon-enough/8677?tag=nl.e539`

2. Vijayan, J.: Researchers use PC to crack encryption for next-gen GSM networks (2010), http://www.computerworld.com/s/article/9144898/Researchers_use_PC_to_crack_encryption_for_next_gen_GSM_networks
3. Flosi, S.L.: comScore Reports (January 2011), U.S. Mobile Subscriber Market Share, http://www.comscore.com/Press_Events/Press_Releases/2011/3comScore_Reports_January_2011_U.S._Mobile_Subscriber_Market_Share
4. Muller, T., Neymayr, T.: Apple Introduces iCloud. Free Cloud Services Beyond Anything Offered to Date (2011), http://www.apple.com/pr/library/2011/06/06Apple-Introduces-iCloud.html
5. Nickinson, P.: A half-million Android devices activated every day, Andy Rubin says (2011), http://www.androidcentral.com/half-million-android-devices-activated-every-day-andy-rubin-says
6. Zorz, Z.: Multiplatform Java botnet spotted in the wild (2011), http://www.net-security.org/malware_news.php?id=1714
7. Maslennikov, D.: Mobile Malware Evolution: An Overview, Part 4 (2011), http://www.securelist.com/en/analysis/204792168
8. Nagy, A.D.: jSMSHider Android Malware Attacks Devices With Custom ROMs (2011), http://pocketnow.com/android/jsmshider-android-malware-attacks-devices-with-custom-roms
9. Dunham, K.: Mobile Malware Attacks and Defense. Syngress Publishing Inc. (2009)
10. http://www.egospodarka.pl/67754,F-Secure-Mobile-Security-7,1,14,1.html
11. http://infonokia.pl/artykuly/symbian/ewolucja-zagrozen-mobilnych
12. Namiestnikow, J.: Ewolucja zagrozen IT w I kwartale, roku (2011), http://www.viruslist.pl/news.html?newsid=663
13. BBC News: New iPhone worm can act like botnet say experts, http://news.bbc.co.uk/2/hi/technology/8373739.stm
14. PandaLabs: New Malware Jumps to 73,000 Samples Every Day, Says PandaLabs, http://www.securityweek.com/new-malware-jumps-73000-samples-every-day-says-pandalabs
15. http://rozmowy.onet.pl/5103,1,sonda_noscript.html
16. McAfee: Mobility and Security. Dazzling Opportunities, Profound Challenges, http://www.mcafee.com/mobilesecurityreport
17. Mick, J.: Reports: Hackers Use Stolen RSA Information to Hack Lockheed Martin, http://www.dailytech.com/Reports+Hackers+Use+Stolen+RSA+Information+to+Hack+Lockheed+Martin/article21757.htm
18. http://www.kaspersky.com/mobile_downloads

Chapter 20
Quantum E-Voting Cards

Marcin Sobota and Adrian Kapczynski

Abstract. In the article the concept of quantum e-voting cards was presented.

20.1 Introduction

In recent years, quantum cryptography has become increasingly popular. This is not just the field of knowledge, which is available only for a narrow group of researchers. It is very often that it uses its advances wherever ensuring an adequate level of data security is crucial. There is several companies in the world which offer devices with action is based on the quantum protocols. Possibility of practical use is one of the reasons to find a new apply of quantum protocols. This article presents conception of quantum e-voting cards.

20.2 Quantum Protocols

In the quantum protocols of key agreement, to encode the data we use the polarization of photons. Appropriate polarization shall be obtained by passing not polarised beam by polarizer oriented according to the need. Action of the polarisator consist of passing photons of polarization compatible with setting polarisator.

Reading of photons polarization is taking place by the passing beam of photons by certain anisotropic materials, such as crystals of calcite or boron nitride and submitting it to the detectors. This kind of materials have the property that if set properly, they pass photons without changing their polarization, while in the case of an incorrect setting of crystal they may – randomly - occurs change of polarization.

Marcin Sobota · Adrian Kapczynski
Academy of Business in Dabrowa Gornicza, Department of Computer Science
ul. Cieplaka 1c, 41-300 Dabrowa Gornicza, Poland
e-mail: {msobota,akapczynski}@wsb.edu.pl
http://www.wsb.edu.pl

A. Kapczynski et al. (Eds.): Internet - Technical Develop. & Appli. 2, AISC 118, pp. 227–233.
springerlink.com

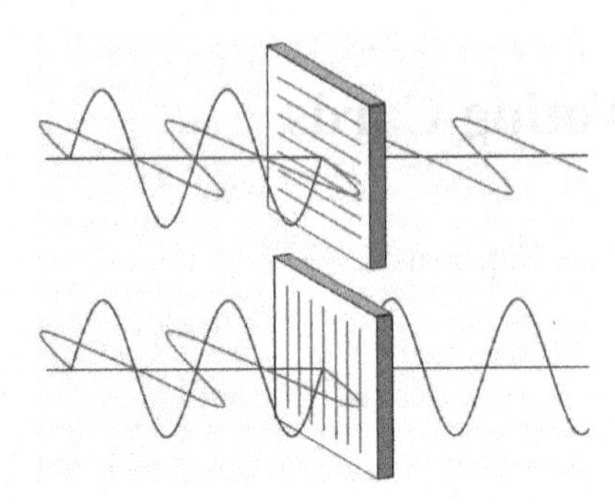

Fig. 20.1 Functioning of light polarization

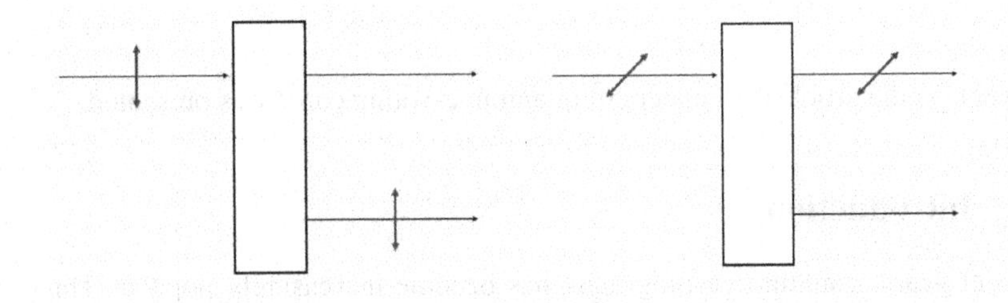

Fig. 20.2 Behavior of photon passing through properly set up center birefringence

Fig. 20.3 Alphabets used in protocol BB84

20.3 Protocol BB84

The idea used in this Protocol has been developed already in the 1970s, but it was completed in 1984. The authors of the Protocol are: Charles Bennett and Gilles Brassard [1] [2] [3]. It is based on the application of the two alphabets:

1. Simple containing photons of polarizations: $0\,^{\circ}$ and $90\,^{\circ}$ (binary 0 and 1),
2. Diagonal - containing photons of polarizations: 45° and 135° (binary 0 and 1).

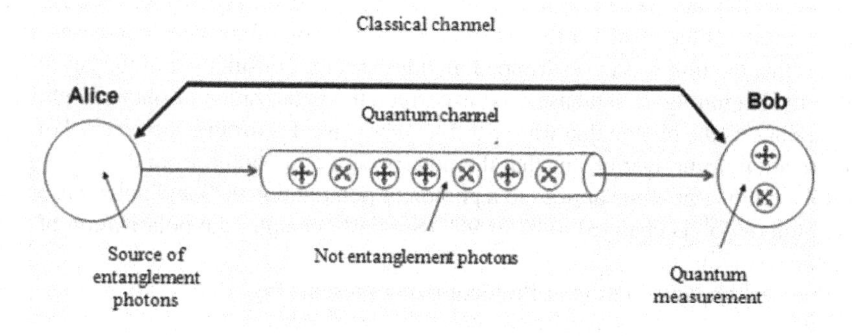

Fig. 20.4 Quantum key distribution without using tangled states

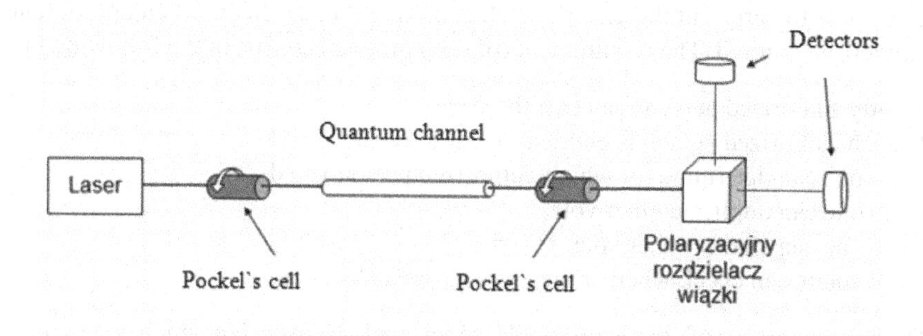

Fig. 20.5 Draft of practical realization of protocol BB84

Agreement of the cryptographic key take place in the following steps:

1. Alice (sender) selects at random one of the four possible polarization and sends to Bob (receiver) the photons of such a polarization. A string of photons is the string of zeros and ones with two quantum alphabets.
2. Bob randomly chooses database straight or diagonal and carries out the measurement of polarization of each photon, that he received from Alice.
3. Bob writes down the results of the measurement keeping them in secret.
4. Bob publicly inform CityAlice which databases he used to measure, and Alice inform him whether selected randomly types of databases were appropriate or not.
5. Alice and Bob store the results of the measurements, for which Bob used proper database. The results of these measurements can be saved in binary form and the obtained string may be used as a cryptographic key.

This way of the exchange encryption key allows you to detect eavesdropping on the link. This is the right of quantum mechanics, according to with passive eavesdropping is not possible. Each eavesdropping is active and introduces false reading in the transmission. To detect eavesdropping we select a certain string of photons, on which Bob and Alice put on the same database and verify whether they got the same results. If the results differ, despite the imposition of the same databases, this means that the link was eavesdropped. In this case, the consultation of the key starts from the beginning. If database used is correct, the polarization of obtained photon is the same as the photon that was sent. If database used is invalid, the reversal of the polarization occur, that means that the polarization of a photon from 0^0 is changed to 45^0 or 135^0 (the same applies to a photon of polarization 90^0) and polarization of a photon of 45^0 is changed to 0^0 or 90^0 (Similarly changes the polarization of the photon is 135^0).

Practical implementation of Protocol BB84 presents Fig. 19.5.

20.4 E-voting Protocols

To be able to carry out the electronic election (e.g. via the Internet) an appropriate protocol is required. The requirements of such protocol consist of the following [4]:

1. only authorized persons can cast the votes,
2. each authorized person is entitled to cast only one vote,
3. no one can determine for whom authorized person voted,
4. no one can duplicate other vote,
5. no one can change other vote,
6. all voters can check whether their votes were calculated.

There are a number of simple protocols, which could be used, but which do not meet all these conditions. Protocol A [1] is a first example.

(Protocol A)

1. each voter encrypts his vote with his private key,
2. in addition, it encrypts vote using the public key of Central Election Commission (CEC),
3. each voter shall send their vote to CEC,
4. CEC decipher the votes, verifies the signatures, calculate the votes and announce results.

Protocol satisfies five of the six conditions, only condition number 3 is not fulfilled. Since the digital signatures of voters are used the CEC would be technically able to check who the person voted.

This restriction does not apply to the Protocol B which is described below [5]:

(Protocol B)

1. each eligible voter receives registration number from the Permission Central Agency (PCA). In addition, PCA stores all issued numbers in order the check that the person does not attempt to vote for a second time,
2. PCA sends the registration numbers to CEC,
3. a voter encrypts its vote and a registration number using public key of CEC,

4. CEC deciphers the package using the private key. Then check whether the registration number is on the list received from PCA. If the number is not present on the list, the votes shall be rejected and in other case the voice is counted, and the registration number is removed from the list,
5. after receiving all the votes CEC announces the results of the election and publish a list of registration numbers and the names of chosen candidates.

This protocol meets all established requirements. The only problem is that the PCA may contact the CEC in order to link the registration number with a name of person who has received that number, and later it could be checked, for whom the given person voted. There are similar protocols in which the PCA and CEC is one institution. Then identification of such person becomes even easier. Nevertheless, it seems that the minutes of such structures are a reasonable compromise between complexity and security. If the only issue is the cover the personal data of a person who receives a registration number, than one can apply the protocol C.

(Protocol C)

1. each entitled person shall be reported to CEC to obtain a registration number. CEC issues such a number and records it in order to verify later. The number is not assigned to a person, while the personal data of a voter that has received a number, are removed from the list of people entitled to vote,
2. voter casting its vote completes the vote with received the registration number and uses CEC public key to encrypt the message,
3. CEC deciphers the message, check whether the vote contained in the registration number is entitled to voting and based on that, the vote is accepted or rejected,
4. after the end of voting, CEC announces results of the elections, together with a list of registration numbers and votes assigned with them.

On the one hand such an approach will result in that the CEC has only registration numbers, which will entitle the holders to vote without the possibility of their assignment to the particular person, and on the other hand, voters may (after the announcement of election results) check whether their vote is correct.

20.5 Conception of Quantum E-voting Cards

Presented in previous chapter protocols of voting includes one important point. Voters need to get cards with names of candidates on which they choose one or many of candidates. In classical voting CEC make one voting card. In quantum e-voting cards CEC have to prepare the same number of cards as is the number of candidates. For example if number of candidates is 100, number of cards is also100. Each candidate have to be placed on every position of the list one time. Additional CEC allots numerical code to each card. Numerical code may be for example 10 bits long (important is that each code have to have the exact length).

Sides use quantum protocol BB84 to set voting card. Besides numerical code, voters have to know what kind of base is used by CEC to encode each bit

Numerical code 1001011010l	Numerical code 1001011010	Numerical code 0100110111
1) XXXXXXXXXX	1) ZZZZZZZZZZ	1) YYYYYYYYYY
2) YYYYYYYYYY	2) XXXXXXXXXX	2) ZZZZZZZZZZ
3) ZZZZZZZZZZ	3) YYYYYYYYYY	3) XXXXXXXXXX
4)........	4)........	4)........
5)........	5)........	5)........

Fig. 20.6 Demonstration e-voting cards

Check sequence	Numerical code	Numerical code	Numerical code	Numerical code	Check sequence

Fig. 20.7 Sequence sending by CEC to voter

of numerical code. This knowledge is important to find correct values by voters (using correct bases). In the day of voting the CEC sent to voters (in form of suitable polarized photons) sequence includes check sequences and numerical codes. Check sequences serves to find potential eavesdropping. Numerical code serves to set voting card.

Voter use random bases to find bits of checking sequence. Next he uses all known bases for each numerical codes with possibility to be used. After that voter one more time uses random bases to find bits of final checking sequence. The idea is very simple. Voter have to find one of known numerical code and this is number of voting card which he have to use to vote. Additionally he checks with CEC (by checking sequences) if communications was eavesdropped (as in BB84 protocol). If channel wasn‘t eavesdropped voter has voting card and can vote. Because only voter and CEC know the number of voting card, voter can vote by using open channel. And this is a strength of using quantum e-voting cards. After set voting card, voter sends to CEC information that he voted for candidate number 57, 68 and 98. Because CEC and voter know which voting card was used, they know which candidate was chosen. Eavesdropper know only number of chosen candidate, but he doesn’t have any information about the name of candidate (number 57 on the list could be any candidate).

20.6 Conclusion

In this article authors presented the conception of quantum e-voting cards from a theoretical point of view and further, empirical study is the next research goal.

References

1. Bennett, C.H., Brassard, G.: Quantum Cryptography: Public Key Distribution and Coin Tossing. In: Proceedings of IEEE International Conference on Computers Systems and Signal Processing, City Bangalore placecountry-region, India (1984)

2. Schneier, B.: Kryptografia dla praktyków, Wyd. 2. WNT, Warszawa (2002)
3. Sobota, M.: Wykrywanie podsluchu w sieci swiatlowodowej podczas konsultacji klucza szyfrujacego. In: Zeszyty Naukowe Politechniki Slaskiej, Organizacja i Zarzadzanie z.20 cz.1, pp. 133–143. Wydawnictwo Politechniki Slaskiej, Gliwice (2004) PL ISSN 1641-3466
4. Sobota, M.: Bezpieczenstwo wymiany klucza szyfrujźcego z wykorzystaniem wybranych protokolow kwantowych. In: Nowe Technologie w Komputerowych Systemach Zarzadzania, pp. 239–244. Wydawnictwo Komunikacji i Lacznosci, Warszawa (2005) ISBN 83-206-1590-09

Chapter 21
Implementation of the OCTAVE Methodology in Security Risk Management Process for Business Resources

Marek Pyka and Ścibor Sobieski

Abstract. One of the most important factors of real live teleinformatic systems are risk management process. The question that arises is how to implement this process in real and big organizations. Every day, most of technological decision maker and financial decision maker are searching methods for secure assets and safe business. In this chapter authors conduct a discussion concerning methodology that improves information management and protection decision making process. The authors describes OCTAVE (The Operationally Critical Threat, Asset, and Vulnerability Evaluation) using real-life examples and reference to the Polish legal regulations. The purpose of the chapter is to present a methodology, which is successfully being employed in Western-Europe countries, United States of America and presents the possibility of using it in Poland, fitting well into the security policies of many organizations. An example of OCTAVE implementation, for small and medium companies, based on Polish law, shall be presented in this article.

21.1 Introduction

After a period of inhibition of business development caused by the economic crisis, most companies have returned to growth and increase the efficiency of their IT infrastructure. Unfortunately, as is levied before, this development is chaotic and not supported by the analysis of business needs for security technologies or any kind of cost model. In constant aspiration to enlarge their assets, companies pay little attention to the issue of data protection on many levels. Constants grow of flow of

Marek Pyka
Academy of Business, ul. Cieplaka 1c, 41-300 Dabrowa Gornicza, Poland
e-mail: mpyka@wsb.edu.pl

Ścibor Sobieski
Department of Theoretical Physics and Informatics, Faculty of Physics and Applied Informatics, University of Lodz, ul. Pomorska 149/153 , 90-236 Lodz, Poland
e-mail: scibor@uni.lodz.pl

A. Kapczynski et al. (Eds.): Internet - Technical Develop. & Appli. 2, AISC 118, pp. 235–252.
springerlink.com

reports about many different types of electronic crimes that causing severe business losses shows how shortsighted such an approach proves to be. Press releases about the spectacular financial loss of many companies caused by disclosing sensitive information, according to executives realize their business processes on the efficiency and security of IT infrastructure [18].

Unfamiliarity with the security problems by management, often results form the desire to save funds, leads to hasty implementation of cheep security solutions or to resorting to the services of poorly trained administrators. Managers often have to hope that buying the newest tool or any kind of technology will solve their security problems. Few organizations started to evaluate what they are actually trying to protect (and why) from an organizational perspective before selecting solutions. They are often implementing best practices [21, 22], ISO just like ISO/EIC 27001 or other security process. People who works in the field of information security, knowns that security issues are been complex and are rarely to be solved simply by applying a piece of technology. Most security issues are firmly rooted in one or more organizational and business issues.

Before implementing security solutions, we should consider complication of the underlying problems by evaluating our security needs and risks in the context of business. Information security requires far more than the latest tool or technology. Organizations must understand exactly what they are trying to protect — and why — before selecting specific solutions. Security issues are complex and often are rooted in organizational and business concerns. A careful evaluation of security needs and risks in this broader context must precede any security implementation to insure that all the relevant, underlying problems are first uncovered. Proper security solutions should protect the company and eliminate the unnecessary risk related to business activity, but we must remember that any solutions has own cost. The proper cost and security efficiency evaluation should include resources that are to be protected, possible threats and losses, as well as expenses related to unpredictable events. It is important to remember that mixing security with functionality may prove to be very expensive. Perfectly secured resources become virtually useless, while a company that aims at highest functionality cannot be perfectly protected.

Methodologies that improve the risk minimization process allow determining the level of balance. In such cases, the resource protection strategy should go together with the tendency to take a level of accepted risk ("How Much Security Is Enough?"— CERT group [20]) [16, 17]. Companies that commit themselves to conduct the described processes should assume the necessity of organizational and hardware changes. Only such approach to planning resource protection allows reaching the desired level of security. One should also perform periodical surveys of resources, threats and risk, as well as take account for the changing conditions (higher level of threat, hardware failure etc.). In the majority of methodologies, a model approach is assumed ("In-depth Defense", Fig. 21.1) [7]. In-depth analysis of the companies architecture lowers the risk of a successful attack and, therefore, data loss. Failing to comply with proper security principles might bring severe business and legal effects. All information in organizations can be divided into two categories: information that a company wants to protect and information that a

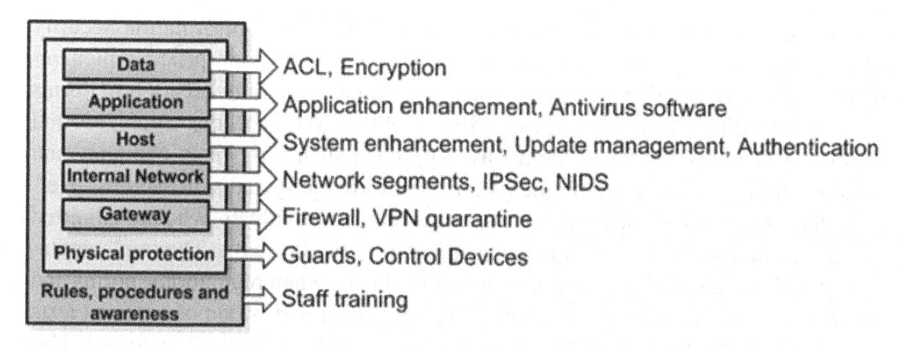

Fig. 21.1 "In-depth Defence" model [7]

company must protect. The Polish law forces the organizations to protect various kinds of information — mainly applies to personal data protection [12]. Some companies must protect information, which is subject to additional legal acts: Personal Data Protection Act [12], Confidential Information Protection Act [13], Banking Confidentiality, Unfair Competition Act [11]. All employees have to comply with these regulations, which also force appropriate security measures, adapted to the particular kind of information that requires protection. Currently there are several dozens of information types in Poland — legal regulations force organizations to familiarize themselves with those acts, as well as to deploy them properly in practice [14]. The administrators in Poland have big problem to implement security process and stay compliant with many different Acts. Because of the varieties and limitations of current security evaluation methods, one may become confused when trying to select an appropriate method for evaluating information security risks in organization. Most of the current methods are "bottom-up": they start with the computing infrastructure and focus on the technological vulnerabilities without considering the risks to the organization's mission and business objectives. Let us know, a better alternative is to look at the organization itself and identify what needs to be protected, determine why it is at risk, and develop solutions requiring both technology — and practice-based solutions. A comprehensive information security risk evaluation approach:

- Incorporates assets, threats, and vulnerabilities.
- Enables decision makers to develop relative priorities based on what is important to the organization.
- Incorporates organizational issues related to how people use the computing infrastructure to meet the business objectives of the organization.
- Incorporates technological issues related to the configuration of the computing infrastructure.
- Should be a flexible method that can be uniquely tailored to each organization.

Implementation of each piece of technology to business processes brings new threats and vulnerabilities that must be analyzed and minimized. Because of these, many security management methodologies has been developed. Currently, one of the widespread information management policies is TISM (Total Information Security Management) [19]. TISM methodology helps to develop an appropriate implementation plan of information resources protection, based on binding acts of law, norms (i.e. PN ISO/IEC 27001:2005, PN ISO/IEC 17799-2:2005) accepted standards and business principles of a given company. Approach such this, given company can establish proper security requirements for teleinformatic systems, as well as determine the tasks and responsibility levels for particular individuals, who are admitted to the corporate secrets and to the management of critical systems. The implementation of TISM methodology includes a theoretical part, when regulations, norms and other security documentation is created and a practical part, when penetration tests and audits are (LAN, Internet/LAN connection, servers, clients) conducted. Unfortunately this methodology is very time-consuming and requires large resources, moreover TIMS methodolody rather fit to US Government requirements, this is the main reason for Polish companies to seek an alternative procedure.

OCTAVE (Operationally Critical Threat, Asset and Vulnerability Evaluation) is a methodology [9] that has not yet taken hold in Poland.

21.2 The OCTAVE Methodology

21.2.1 An Introduction to the OCTAVE Methodology

One way to create a context-sensitive evaluation approach is to define a basic set of requirements for the evaluation and then develop a series, or family, of methods that meet those requirements. Each method within the approach could be targeted to a unique operational environment or situation.

OCTAVE [9] is a methodology that improves the decision making process concerning protection and management of resources in a company. It was developed in year 2001 by the Carnegie Mellon University. Risk assessment is based on three basic principles of security administration: confidentiality, integrity, availability — by means of simple classification of critical information, one will receive a plan of protection for the given information. Authors conceived the OCTAVE project to define a systematic, organization-wide approach to evaluating information security risks comprising multiple methods consistent with the approach. The OCTAVE designed the approach to be self-directed, enabling people to learn about security issues and improve their organization's security posture without unnecessary reliance on outside experts and vendors. An evaluation by itself only provides a direction for an organization's information security activities. Meaningful improvement will not occur unless the organization follows through by implementing the results of the evaluation and managing its information security risks. OCTAVE is an important first step in approaching information security risk management. The OCTAVE approach is defined in a set of criteria that includes principles, attributes, and outputs. Principles are the fundamental concepts driving the nature of the evaluation. Its define the philosophy that shapes the evaluation process. For example, self-direction is one of the principles of OCTAVE. The concept of self-direction means that people inside the organization are in the best position to lead the evaluation and make decisions. The requirements of the evaluation are embodied in the attributes and outputs. Attributes are the distinctive qualities, or characteristics, of the evaluation. They are the requirements that define the basic elements of the OCTAVE approach and define what is necessary to make the evaluation a success from both the process and organizational perspectives. Attributes are derived from the OCTAVE principles. For example, one of the attributes of OCTAVE is that the analysis team should be an interdisciplinary team staffed by personnel from the organization leads the evaluation. The principle behind the creation of an analysis team is self-direction. Finally, outputs define the outcomes that an analysis team must achieve during the evaluation. Table 1 lists the structure of the principles, attributes, and outputs that we will examine in this chapter. We begin our exploration of the OCTAVE approach in the next section by looking at principles.

Tab.1. Information Security Principles, Attributes, and Outputs [9].

Information Security Principles, Attributes, and Outputs Principles Attributes Self-direction Adaptable measures Defined process Foundation for a continuous process Foward-looking view Focus on the critical few Integrated Management Open communication Global perspective Teamwork Analysis team and augmenting analysis team skills Catalog of practices Generic threat profile Catalog of vulnerabilities Defined evaluation activities Documented evaluation results Evaluation scope and next steps Focus on risk and focused activities Organizational and technological issues Business and information technology participation Senior management participation and collaborative approach Outputs Phase 1 Phase 2 Phase 3 Critical assets Security requirements for critical assets Threats to critical assets Current security practices Current organizational vulnerabilities Key components Current technology vulnerabilities Risks to critical assets Risk measures Protection strategy Risk mitigation plans.

OCTAVE recommends to organize appropriate workshops in a company. During those workshops, the employees themselves make the decisions concerning the level of importance for particular data resources. In order to make the best decision possible, it is necessary to attribute threat categories to particular resources. By resources we mean information as well as systems that process the information (along with supplementary applications), the company employees themselves are also considered as a resource. In OCTAVE, threats are described by three logical structures — a threat profile is created by proper reference to those structures. Thanks to this, the OCTAVE implementation team is able to determine the threat range for the given resources. If the team determines the level of risk and understands the potential damage due to data loss, it will be able to take proper steps to minimize it and ensure better protection of the organization's resources.

21.2.2 Characteristics of OCTAVE

The OCTAVE methodology is interesting, because it is perfectly adjusted to the policies of many companies. Every company has information that requires protection. Mere knowledge of the risks will not protect our business. OCTAVE should be applied regularly, because in large companies, the flow of information is constant. It is common for some given data to gain more importance over time. The lack of regularity may lead to data compromise or legal consequences. OCTAVE is also time and resouruce consuming process (like TISM), however, it should not be neglected, but is independent vendor and more elastic then TISM. For organizational purposes, it is divided into three phases [9] (see Fig. 21.2.).

Phase 1. **Build Asset-Based Threat Profiles** — it involves the evaluation of the company's security strategy and the determination of possessed resources. During this phase, the employees should be aware of the resources possessed by the company and which of them require special protection. Security requirements for this type of resources should be determined. The employees describe the security measures applied by the company so far and attempt to determine weaknesses in this strategy. This phase involves gathering introductory information; it is mainly based on the interviews with the employees. Phase 1 makes the staff aware of the importance of data protection, as well as describes the potential losses that could emerge in case of a vital data loss.

Phase 2. **Identify Infrastructure Vulnerabilities** — involves the evaluation of the information management system. It is mainly based on the data gathered during Phase 1. Data protection vulnerabilities are being surveyed with focus on technological issues. Key issues for the future strategy are being determined. This phase involves gathering data from the employees of the IT department, executives and other staff. A common solution should be developed, without hindering the present business model of the company.

Phase 3. **Develop Security Strategy and Plans** — this is the phase of risk analysis. Information gathered in Phase 1 and Phase 2 are used to assess the risk of data compromise in the company and the risk associated with the company's business activity. The security strategy and ways of minimizing the risk of data loss are being developed. With the exact information concerning the business model of the company, we are able to determine the attack types, which might take place in the future. In phase 3, the exact procedures are being created. The team that conducts the OCTAVE process must verify the legal status of the guidelines.

The line of business of the given organization plays a key role in the final decisions concerning data management strategy. OCTAVE assumes that the level of risk tolerance for a given company should be determined first — it's similar to regular risk management. The security strategy should be developed later, in order to ensure the highest possible level of protection, without hindering business activities. Issues taken into consideration are: legal limitations, expenses of implementing the new strategy, productivity and security of consumer data. The documentation of OCTAVE methodology is very extensive, which ensures detailed data verification. However, an in-depth analysis of resources is not always necessary. For such cases, OCTAVE-S has been developed [10]. It is used in small or medium-sized companies (employing not more than 100 people). The differences between those processes are visible during the conduction of workshops for the employees. Only several people are needed to perform the risk analysis. It is said that OCTAVE-S is a much easier

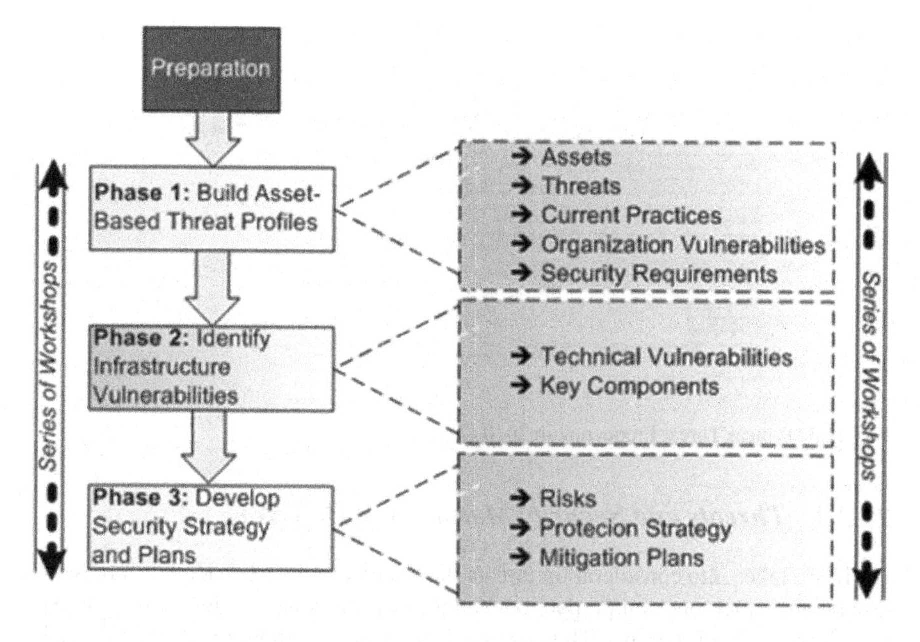

Fig. 21.2 The division of the OCTAVE process with sub-processes. (OCTAVE manual)

solution. However, the choice of the variant depends only on the needs of an organization. There are companies, where both OCTAVE and OCTAVE-S would turn out to be a good solution — in such cases an initial calculation may prove to be helpful in determining the optimal methodology. The conduction of both processes takes about 3 days of time. However, when developing a strategy with the aid of OCTAVE, one must consider the budget available for this purpose.

If the OCTAVE methodology is to bring the intended results, the company must be aware of the threats that might occur for the possessed resources.

Fig. 21.3 illustrates the results of a survey conducted by CERT Polska in 2010 [2]. Attack on information security takes up only 0.74% (0.39% in 2008 year) of all threats, while Information gathering goes as high as 9.79% (4.79% in 2008). From the same report one can learn that commercial companies are the most common attacks are computer frauds about 36.5%, malware 13.5% and Illegal content 28.93%. It is easy to draw a conclusion that the probability of Computer fraud, Illegal content and Mallware taking place in a company, is rising proportionally with the size of the company and the amount of information possessed by it. CERT Polska noted 12 027 544 of such incidents in 2010, however, this is only the number of officially and automatically reported cases. Such statistics should make companies aware of the importance of protecting their resources. The OCTAVE methodology is very elastic in implementation. Basing on such reports, one can rule out the most likely threats and focus on finding solutions to the less common ones.

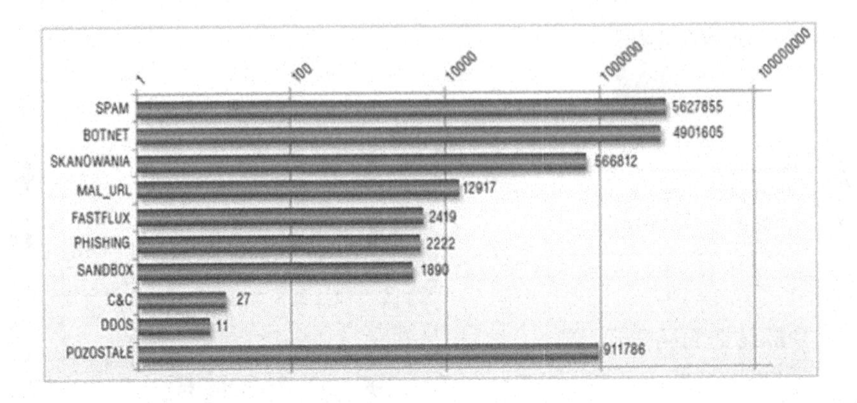

Fig. 21.3 Typical Threat Frequency in 2010 [2].

21.2.3 Threats and Security Measures in OCTAVE

OCTAVE takes into consideration threats illustrated on Fig. 21.4. Higher forces are threats, for which one can prepare but cannot directly influence them (for example fire, flood, network failure). The described methodology should minimize the risk resulting from improperly organized data processing (Organizational Negligence). Threats associated with this issue are, for example, unauthorized access to resources, privilege escalation or lack of resource control — these are the most commonly omitted aspects of security in many organizations. Human errors are also highly important, they make over 70% of threats. Workshops offered by OCTAVE also make the staff more aware of various threats. Technological errors may sometimes directly result from deliberate actions. Deliberately causing a network failure becomes a technological error and a direct trouble for the company. Dishonest staff or external attacks are usually taken into consideration in a later stage. In response to the threats defined in the preliminary phase, OCTAVE also includes proposing certain solutions, in order to minimize the likelihood of those threats to occur. Security

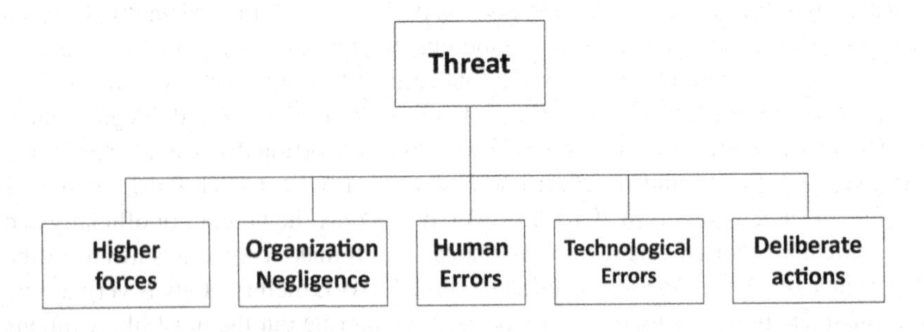

Fig. 21.4 The division of existing threats — Federal Office for Information Technology Security (BSI) [1].

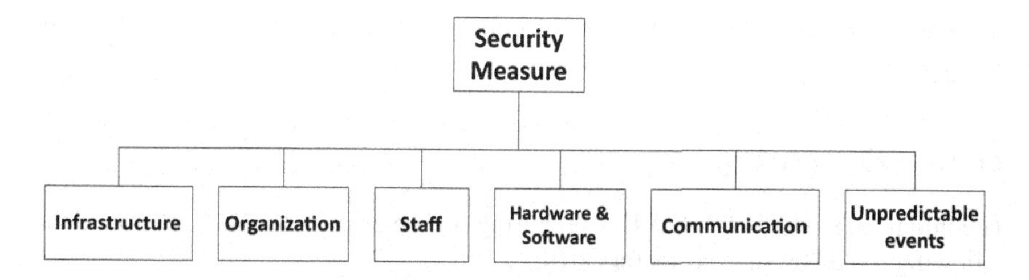

Fig. 21.5 The logical division of security measures in the OCTAVE process [1].

solutions for threats named in Fig. 21.5 are being considered. Those are general assumptions; however, drawing up details for them might result in a very good starting point for resource protection. The security of a company involves various factors, which are directly adapted to business purposes or attack susceptibility. The conduction of OCTAVE methodology allows us to notice certain details, which may help to determine those factors. Let us remember that there is no such thing as a perfect security. It's well known, there is no plan in a real world to eliminate all risks. There are threats which we are unable to predict. However, OCTAVE is not used to create emergency plans for such cases; it is a methodology that helps to determine the risk, not to counteract existing threats.

OCTAVE is useful in preventing losses, not in repairing damage. One can only calculate the risk of a given event to occur — this is performed thanks to information gathered during phases 1 and 2. Emergency plans are being developed, which are to be applied in case of an unpredicted event. The security improvement strategy based on the OCTAVE methodology is a good guideline for a company and its employees — it tells them what to do, in order to minimize the risk of losses. Thanks to this, the company gains a reputation of being trustworthy and learns how to manage information security.

21.3 OCTAVE Workshops

The OCTAVE methodology involves two types of workshops:

1. facilitated discussions with various members of the organization;
2. workshops in which the analysis team conducts a series of activities on its own.

All workshops have a leader and a scribe. The leader is responsible for guiding all workshop activities and ensuring that all of these (including preparatory and follow-up activities) are completed. The leader is also responsible for ensuring that all participants understand their roles and that any new or supplementary analysis team members are ready to participate actively in the workshop. All workshop leaders should also make sure that they select a decision-making approach (e.g., majority vote, consensus) to be used during the workshops. Scribes are responsible for recording information generated during the workshops, either electronically or on paper. Please note that you might not have the same leader or scribe for all workshops.

21.3.1 Preparation

The initial focus of the OCTAVE is preparing for the evaluation. We have found the following to be key success factors [10]:

- Getting senior management sponsorship. This is the top success factor for information security risk evaluations. If senior managers do not support the process, staff support for the evaluation will dissipate quickly.
- Selecting the analysis team. The analysis team is responsible for managing the process and analyzing information. The members of the team need to have sufficient skills and training to lead the evaluation and to know when to augment their knowledge and skills by including additional people for one or more activities.
- Setting the appropriate scope of the OCTAVE. The evaluation should include important operational areas, but the scope cannot get too big. If it is too broad, it will be difficult for the analysis team to analyze all of the information. If the scope of the evaluation is too small, the results may not be as meaningful as they should be.
- Selecting participants. During the knowledge elicitation workshops (processes 1 to 3), staff members from multiple organizational levels will contribute their knowledge about the organization. They should be assigned to workshops because of their knowledge and skills, not solely based on who is available.

The goal of preparation is to make sure that the evaluation is scoped properly, that the organization's senior managers support it, and that everyone participating in the process understands his or her role.

21.3.2 Phase 1: Build Asset-Based Threat Profiles

In phase 1 you begin to build the organizational view of OCTAVE by focusing on the people in the organization (see Fig. 21.6.).

Processes 1 to 3. The analysis team facilitates knowledge elicitation workshops during processes 1 to 3. Participants from across the organization contribute their unique perspectives about what is important to the organization (assets) and how well those assets are being protected. The following list highlights the audience for each of the processes:

Process 1. **Identify Senior Management Knowledge.** The participants in this process are the organization's senior managers.

Process 2. **Identify Operational Area Management Knowledge.** The participants in this process are the organization's operational area (middle) managers.

Process 3. **Identify Staff Knowledge.** The participants in this process are the organization's staff members. Information technology staff members normally participate in a separate workshop from the one attended by general staff members.

Four activities are undertaken to elicit knowledge from workshop participants during processes 1 to 3. This is the identification of: assets, relative priorities, areas of concern, security requirements for the most important assets and capture of knowledge of current security practices and organizational vulnerabilities.
Process 4. **Create Threat Profiles.** The participants in this process are the analysis team members. During process 4, the team identifies the assets that are most critical to the organization and describes how those assets are threatened. Process 4 comprises the following activities: consolidating information from processes 1 to 3, selecting critical assets, refining security requirements for critical assets, identifying threats to critical assets.

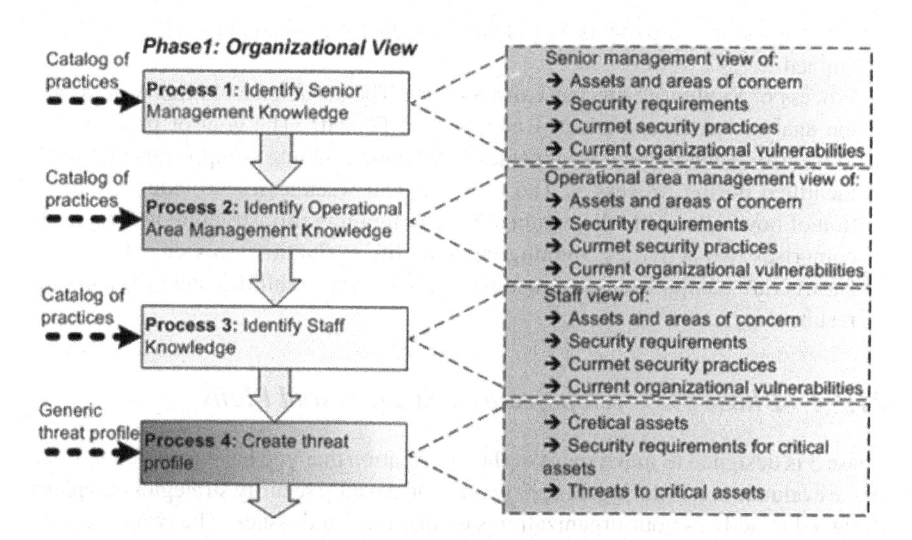

Fig. 21.6 Phase 1: Build Asset-Based Threat Profile.

21.3.3 Phase 2: Identify Infrastructure Vulnerabilities

Phase 2 is also called the "technological view" of the OCTAVE, because this is where you turn your attention to your organization's computing infrastructure. The second phase of the evaluation includes two processes (see Fig. 21.7).

Process 5. **Identify Key Components.** The participants in this process are the analysis team and selected members of IT staff. The ultimate objective of process 5 is to select infrastructure components to be examined for technological weaknesses during process 6. Process 5 consists of two activities: identifying

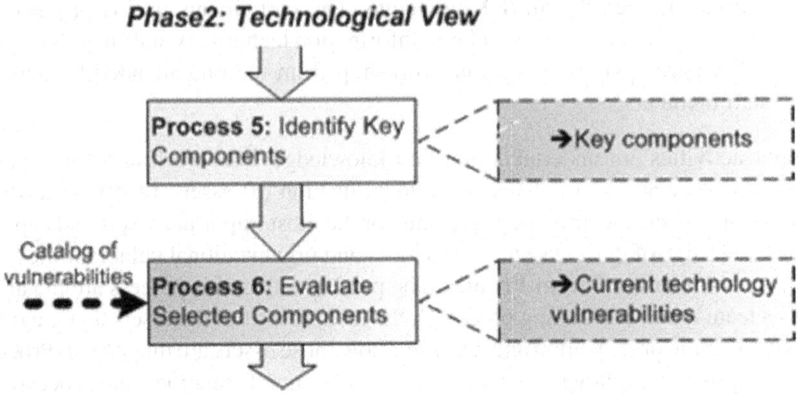

Fig. 21.7 Phase 2: Identify Technological Vulnerabilities.

key classes of components and identifying infrastructure components to be examined.

Process 6. **Evaluate Selected Components.** The participants in this process are the analysis team and selected members of IT staff. The goal of process 6 is to identify technological weaknesses in the infrastructure components that were identified during process 5. The technological weaknesses provide an indication of how vulnerable the organization's computing infrastructure is. Process 6 comprises two activities: running vulnerability evaluation tools on selected infrastructure components, reviewing technology vulnerabilities and summarizing results.

21.3.4 Phase 3: Develop Security Strategy and Plans

Phase 3 is designed to make sense of the information that you have gathered thus far in the evaluation. It is during this phase that you develop security strategies and plans designed to address your organization's unique risks and issues. The two processes of phase 3 are shown in Figure 21.8.

Process 7. **Conduct Risk Analysis.** The participants in process 7 are the analysis team members, and the goal of the process is to identify and analyze risks to the organization's critical assets. Process 7 includes the following three activities: identifying the impact of threats to critical assets, creating risk evaluation criteria, evaluating the impact of threats to critical assets.

Process 8. **Develop Protection Strategy.** Process 8 includes two workshops. The participants in the first workshop for process 8 are the analysis team members and selected members of the organization (if the analysis team decides to supplement its skills and experience for protection strategy development). The goal of process 8 is to develop a protection strategy for the organization, mitigation plans for the risks to the critical assets, and an action list of near-term actions.

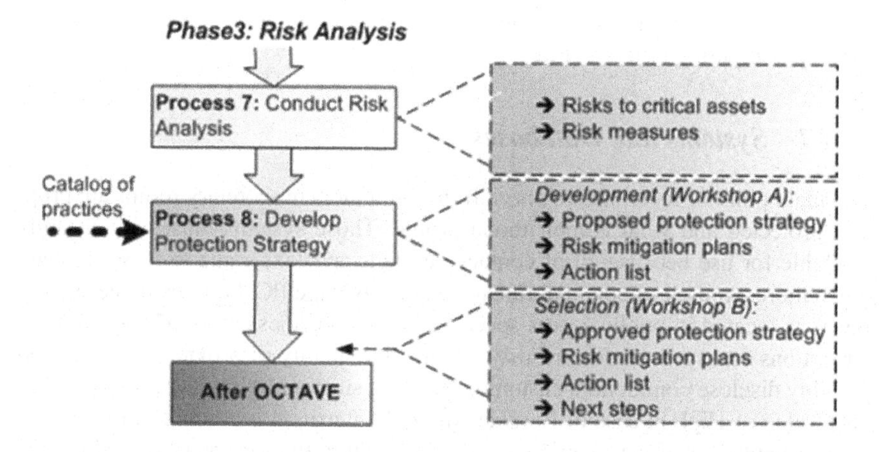

Fig. 21.8 Phase 3: Develop Security Strategy and Plans.

The following are the activities of the first workshop of process 8: consolidation of information from processes 1 to 3, review of risk information, creation of protection strategy, mitigation plans and action list. In the second workshop of process 8, the analysis team presents the proposed protection strategy, mitigation plans, and action list to senior managers in the organization. The senior managers review and revise the strategy and plans as necessary and then decide how the organization will build on the results of the evaluation. The following are the activities of the second workshop of process 8: preparation for a meeting with senior management, presentation of risk information, review and refinement of protection strategy, mitigation plans, and action list, creation of next steps. At this point, the organization has completed the OCTAVE methodology.

21.4 Example of OCTAVE Implementation

This case analysis is a result of conducting the OCTAVE process in a low and financial institution in Poland, which is a medium sized organization. The main area of business is counseling on fundraising and grants from the European Union and Polish government. One of the main business assumptions of the described company was to protect the personal and business data of the clients. Taking this into consideration, we had to take account for following legal acts that are binding in Poland: Personal Data Protection Act [12] and Confidential Information Protection Act [13], norms: ISO/IEC 27001:2005, PN ISO/IEC 27002:2007 [14] and Confidential Information (formulation derived from the Unfair Competition Act [11]). The company also was interested in securing the computers operated by consultants, business analytics, lawyers and other staff. Consultants and analytics had access to confidential clients data and scope of projects — available documentation includes sensitive information for obtaining higher financial subsidies, and very often include "know-how" of companies. The company had doubts about the legal status of their existing computing infrastructure.

21.4.1 Systems and Databases

The main point of focus was to check if the computers in secretary rooms are properly protected and what is their threat profile. Those systems must be constantly available for use because all of contacts, e-mails and faxs came to it, so they are much more exposed to potential threats than an average PC. Of course, these computers is located in public access spaces. From the business point of view, any interruptions or errors would seriously threaten the quality of provided services and possibly disclose confidential customer data. The surveyed systems reached the Impact Value of High in case of Loss/destruction. Interruption of ability of operation is unacceptable, however, it might occur. Such extraordinary events stay within the boundaries of tolerance for this organization. The results of research were compared with the company's policy and gained acceptance.

The database that stores information about clients may be subject to unauthorized access and illegal personal information gathering. Such a crime results from improper protection of computers that have access to the database, the server on which the database is stored and finally, the database itself. This threat was marked as Medium, since it does not expose the company to direct financial losses. Modifications of the database resulting from employee negligence are important, however, their overall impact on the business process is Low, because the can be easily rolled back. The company's database backup systems was found to be outdated. In case of a system failure, restoring date from this backup system will not be done with successful or would it be very long process. Outdated procedures that lead time to restore the system after the crash lasted two times longer than would result from its complexity.

21.4.2 Personal Computers and Laptops

In many organizations internet access is a key issue in planning security structure, so in this case we have the same situation. The organization has a remote office and few employers working from Internet. All personal computers (operated by analytic and other staff) have internet access. Analysis process was determined that they are not a vital tool of work, but seriously improve the business process, when present. In research one was took account of such factors as: disclosure, modification, loss or destruction and unavailability of information on personal computers. Since the organization operates on 12/6 basis, also its personal computer should be able to operate constantly and the staff should have permanent access to them. This can be achieved by employing the basic rules of security, such as using sufficiently strong passwords and mutual authentication on laptops. Loss/destruction of PCs has a critical meaning for the organization, since it could lead to an interruption of the business process together with financial losses. The same applies to Interruption and Modification, particularly in relation to applications that support the operation of this systems.

21.4.3 Documentation

The documentation section includes both the documents related with the operation of the company itself and documents related with the clients, for example, subject of an application to the grant. Documentation should be available in electronic and paper form. There should be permanent access to the documents and they are to be stored in a logical structure. There're able to observe a tendency to store documents on desks or shelves. This should never happen — important documents are not to be placed in easy to spot locations. Only staff that runs the documentation or requires it for their work should have access to the documents. Any change in clients' files may directly affect their business process, which is unacceptable. Analysis showed also the possibility of complete destruction of the documents (fire, water etc.). In phase one, according to the OCTAVE methodology, organization must determined the main assets of the company, as well as their access paths and analyzed in detail the access to documentation, PCs, databases (of clients and staff) and systems running dedicated applications enhancing the business development.

The biggest problems were the PCs — its were poorly or not at all protected. Moreover the documentation storage proved to be a troublesome area as well; in this case was suggested a separate room, or a place in a room, which only a limited number of employees could access. As far as network structure is concerned, was suggested the purchase of a quality two way software or hardware firewall, which should eliminate the risk of viruses or any sort of malware penetrating the network (at least to the Medium level). Further research revealed that computers often lacked security updates of operation systems. The company should also consider training its staff in the field of security and risk management; such a training should include basic issues of computer operation with stress on security and making the staff aware of the dangers resulting from internet access. The remaining resources shall not be discussed in this chapter.

21.4.4 Survey

Additional surveys were conducted in the following groups of employees: Senior Managers, Operational Area Managers, Staff, IT Staff, Lawyers, Business Analytics. The majority of departments produced similar answers. The most disturbing answers concerned the lack of a business continuity plan and the fact, that the staff did not understand its role in resource protection. This leads to a situation where people fail to comply with many regulations, because they are not aware of the danger. The IT staff was proven to be poorly trained and was unable to properly train the remaining staff. The lack of proper internet advertising was also noted. The employees claimed that they require better hardware and that it takes time to purchase some, which brings us to the conclusion that the company's budget is not particularly big in this field. The company did not conduct any sort of security audits. All employees claimed that the company did not possess the full documentation required by organizations controlling the legal status of the company. The majority did not know to who they should turn in case of a critical situation. The above conclusions close the phases 2 and 3.

21.4.5 Summary of Presented Example

The Network Infrastructure Map is shown in Figure 21.9. Secretary rooms are also included. Exclamation mark highlights the critical assets of the company. Personal computers (especially containing data of Client Stuff) are themselves the system of interest. The conclusions was drew during the implementation of OCTAVE methodology were based on a 3 level grade: High, Medium, Low [9]. These expressions described the level of influence of a particular phenomenon on the everyday functioning of the company. There're also considered the influence of the clients, mobile workers and the employees' attitude towards the strategy that the company was using before our surveys. The Evaluation Criteria included following areas: sensitive data of customers, customer confidence and reputation, productivity, legal penalties, finances and other.

While the analysis assumes that worked out good practices and security criteria will become the standard for the organization. Presented above solutions is a complete set of processes included in the OCTAVE methodology. On the basis of the workshop report, the company should take all possible steps to immediately eliminate all the High risk level of the threat and Medium level should be eliminated within three months.

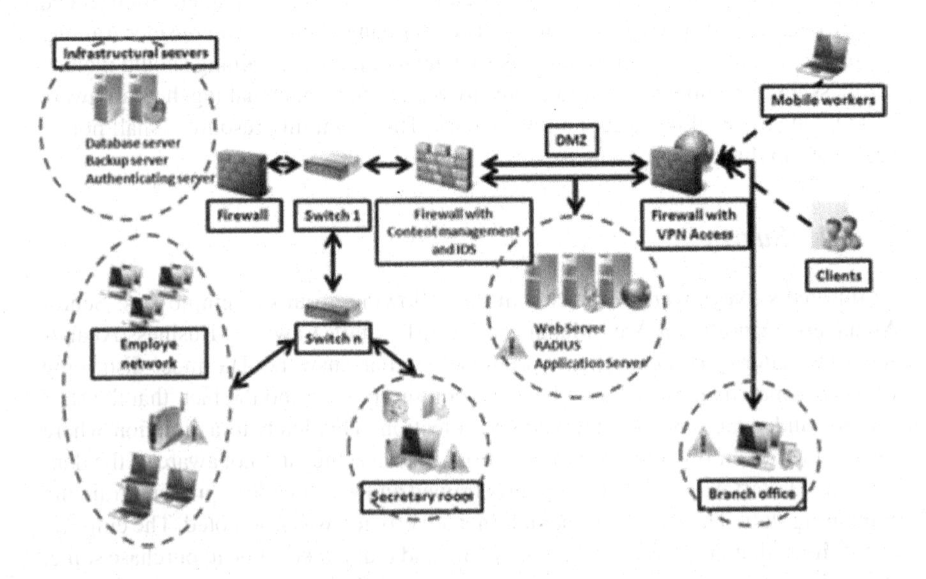

Fig. 21.9 Network Infrastructure Map.

21.5 Summary

Currently a few threat management methodologies exist, including ENSI, TISM, Microsoft MSF, MOF, OCTAVE. Each of the mentioned methodologies puts stress on different elements or forms of resource analysis teamwork organization, but all of them have the same purpose: to improve the security level of our systems and data. The choice of a proper methodology depends on the company-specific factors, but the authors are trying to point out that using OCTAVE allows for developing security principles, which are well adapted to the Polish market and company profile. The OCTAVE approach for self-directed security evaluations was developed at the influential CERTő Coordination Center. This approach is designed to help you:

- Identify and rank key information assets.
- Weigh threats to those assets.
- Analyze vulnerabilities involving both technology and practices.

OCTAVE enables any organization to develop security priorities based on the organization's particular business concerns. The approach provides a coherent framework for aligning security actions with overall objectives. The company workshops, which are described above in detail, are a huge innovation in work organization. Thanks to such approach, all participants feel to be the authors of the policy implementation and bear the moral responsibility for it to some extent. The described example of OCTAVE implementation proves that this methodology is as efficient as other methodologies used commercially these days. The attractive approach to teamwork in this methodology allows the authors to forecast an increase in the amount of its implementations.

References

1. Bialas, A.: Podstawy bezpieczenstwa systemow teleinformatycznych, praca zbiorowa pod red. Bialas A. Wydawnictwo Pracowni Komputerowej Jacka Skalmierskiego, Gliwice (2002) (in polish)
2. CERT Polska — Annual Report 2011, `http://www.cert.pl/PDF/Raport_CP_2010.pdf` (cited July 10, 2011)
3. Council Resolution of February 18, 2003 on a European approach towards a culture of network and information security (2003)
4. Directive 2002/58/EC of the European Parliament and of the Council of July 12, 2002 concerning the processing of personal data and the protection of privacy in the electronic communications sector (Directive on privacy and electronic communications) (2002)
5. Fromholz, J.M.: The European Union Data Privacy Directive, 15 Berkeley Tech. L.J. 471, 472 (2000)
6. Liderman, K.: Podrecznik administratora bezpieczeństwa teleinformatycznego. MIKOM, Warszawa (2003)
7. Łukawiecki, R.: A Holistic View of Enterprise Security. Project Botticelli Ltd.
8. Microsoft Internet Threat Report (November 2010)
9. OCTAVE Method Implementation Guide
10. OCTAVE-S Implementation Guide

11. Polish act of law: Ustawa z dnia 16 kwietnia, r. o zwalczaniu nieuczciwej konkurencji z późniejszymi zmianami (1993) (in polish)
12. Polish act of law: Ustawa z dnia 29 sierpnia, r. o ochronie danych osobowych z późniejszymi zmianami (1997) (in polish)
13. Polish act of law: Ustawa z dnia 22 stycznia, r. o ochronie informacji niejawnych z późniejszymi zmianami (1999) (in polish)
14. Polish norms: PN ISO/IEC 27001:2005, PN ISO/IEC 27002:2007 (2007) (in polish)
15. PWN Oxford Dictionary. Oxford University Press (2004)
16. Saint-Germain, R.: Information security management best practice based on ISO/IEC 17799; the international information security standard provides a framework for ensuring business continuity, maintaining legal compliance, and achieving a competitive edge. Information Management Journal (2005)
17. Schechter, S.E., Smith, M.D.: How Much Security is Enough to Stop a Thief? In: GI 1973. LNCS, vol. 1/1973, pp. 122–137 (2003)
18. Smith, L.M., Smith, J.: Cyber Crimes Aimed at Publicly Traded Companies: Is Stock Price Affected? In: American Accounting Association Southwest Region Annual Meeting, Oklahoma City. Collected Papers and Abstracts, p. 129 (March 2006)
19. TISM (Total Information Security Management) — methodology documentation, version 1.4
20. `https://buildsecurityin.us-cert.gov/daisy/bsi/articles/best-practices/management/566-BSI.html` (cited July 10, 2011)
21. `http://www.sis.pitt.edu/dtipper/2825/ISO_Article.pdf` (cited July 10, 2011)
22. `https://buildsecurityin.us-cert.gov/bsi/articles/best-practices/management/566-BSI.html` (cited July 10, 2011)

Chapter 22
Tools and Methods Used to Ensure Security of Processing and Storing Classified Information in Databases and IT Systems and Their Impact on System Performance

Łukasz Hoppe, Łukasz Wąsek, Arkadiusz Gwóźdź, and Aleksander Nawrat

Abstract. In the article aspects of classified information protection in data bases and IT&T systems were presented. Attention was drawn to formal and legal requirements which in an obligatory manner must be fulfilled by such systems. In the article the issue is discussed of the necessity to obtain system accreditation by the Agency for Internal Security (ABW), which is appropriate in internal security matters of the Republic of Poland. Security measures applied in IT systems are discussed in detail concentrating on physical security measures of IT systems and databases as well as programmary security measures. In the article one could not omit the issue of consequence of a large number of security measures indicating side effects possible to occur.

22.1 Introduction

The dynamic development of information technologies in recent years initiated an era of informatisation in institutions and government agencies which process classified information. Initially, the classified information was processed manually. Documents generated were gathered in rooms prepared for that purpose, generated by a limited group of people and stored in the appropriate secret offices. Along with the development of IT sciences, the needs of workers have evolved to such an extent that that the paper documentation is no longer sufficient. The quantity of documentation produced has increased, the viewing, processing and storage of which is becoming

Łukasz Hoppe · Łukasz Wąsek · Arkadiusz Gwóźdź
WASKO S. A., Berbeckiego 6, 44 - 100 Gliwice, Poland
Silesian University of Technology, Gliwice, Poland

Aleksander Nawrat
Silesian University of Technology, Gliwice, Poland
e-mail: Aleksander.Nawrat@polsl.pl

A. Kapczynski et al. (Eds.): Internet - Technical Develop. & Appli. 2, AISC 118, pp. 253–267.
springerlink.com

more and more challenging. The solution to that have become information systems which currently support the entire processing and storage of classified information.

22.2 Classified Information – Legal Considerations

Appropriate protection is as important as the acquisition and processing of information. The Polish law imposes on persons and institutions handling classified information many responsibilities and conditions that must be met to be able to correctly process and protect classified information processed in ICT systems. The most important legislation governing the processing of classified information and ICT security of such data are the following:

- The law dated 22 January 1999 on the protection of classified information;
- Regulation of the Prime Minister of 25 August on the basic requirements for ICT security;
- Recommendations of ABW (Internal Security Agency)(or BSC) for: physical security, ICT security, electromagnetic protection, development of documentation for the specific requirements of system security, analysis and risk management requirements for software.

The ICT Security is achieved through a set of technical and procedural solutions to mitigate possible threats to an acceptable level. The ICT security concept involves three basic questions:

- What is to be protected? It is the object of protection.
- Against what is it to be protected? It is the identification of hazards.
- How is to be protected? Incorporating the means of protection.

The answer to the first question is contained explicitly in the Law, which contains a definition of protected information that is processed in the ICT systems. This concept has been extended in a document with recommendations on the ICT security of DBTI ABW, according to which also the entire system must be protected, along with devices, in which the classified information and the data itself is processed.

Classified information should be protected from unauthorized access, disclosure, phishing, theft, unauthorized modification, falsification and the possible consequences of those actions. Therefore, it seems important to identify hazards and counteract them appropriately.

The third issue is related to the technical and procedural compliance with the requirements for protection of classified information, which are imposed on the system or telecommunication network, in which the classified information is processed.

According to the applicable law, those requirements are determined by levels: organizational, procedural, technical and personal, to which refer the security service recommendations, related to the protection of classified information processed in the IT systems. All the solutions should be included in the drafting of the document: "plan of protection of classified information" in an organizational unit, as required by the article 18 of the Act.

The main requirements of security of classified information processing environments include:

- Separation of administrative and security zones, in which the classified information will be processed;
- Preparing a security plan;
- Providing continuous access to critical system components;
- Monitoring access to the premises;
- Electromagnetic protection of network devices;
- Preventing monitoring in the network;
- Encryption of database resources and transmission;
- User authentication and authorization;
- Ensuring accountability of user actions;
- Limiting and control of access to data;
- Ensuring reliability of database software;
- Creating backups;
- Providing emergency power supply of key equipment.

According to the provisions of the Act, for the protection of classified information in an organizational unit the manager of that unit is responsible, who assigns the protection agent responsible for ensuring compliance with regulations on the protection of classified information. Special responsibility for the ICT systems take the persons acting as the administrator and IT security inspector (Article 64 of the Act). The last link are the persons, end-users of the ICT system, who processes the classified information, with their responsibilities determined by the Act.

An IT system used to process the classified information must have an accreditation certificate (Article 60 of the Act). Certification of the systems and networks in the civilian sector, in which the classified information is processed, it dealt with by the Department of ICT Security ABW. The basis to obtain the security accreditation certificate is a positive endorsement by ABW of the detailed documents of the System Safety Requirements and Procedures for the Safe Operation, which shall be prepared by the entity applying for the certificate. Prerequisite for the granting of the certificate is also getting the personnel security clearance and credentials to obtain a positive result following an audit.

22.3 Ensuring Security of Classified Information – Tools and Methods

The issue of managing the security of ICT systems that process classified information is a complex subject, often demanding making strategic decisions. Therefore, before attempting to protect the system, a very thorough risk analysis should be carried out. Risks should be identified, faced by information processed in the system, and then, using an appropriate selection of security measures, effective protection against the identified threats can be achieved. This is a difficult task, often impossible to complete due to the complexity and diversity of issues. It should be kept in mind that the level of system security is not constant, it changes over time, and the choice of appropriate safeguards must be subject to continuous updating and modifications. The issues related to risk analysis in terms of protection of classified information have also been listed in the guidelines of DBTI ABW.

The ICT security is the product of the security of infrastructure components. In particular, care must be taken about:

- Physical security;
- Security of transmission;
- Security of devices;
- Security of software;
- Virus protection;
- Authorization and authentication of users.

The ICT Security is ensured by protecting information processed in ICT systems against loss of properties ensuring the security, in particular the loss of: confidentiality, availability and integrity.

While deciding on a specific security application in IT systems, one should focus on both, the physical (technical) safeguards of the information systems and databases and software protections.

22.4 Physical Security

The physical protection of systems involves placing a device or a communication data system in the safety zone, and the application of measures to ensure protection against unauthorized access, viewing and eavesdropping. The physical security is implemented by using technical means such as: access control systems, alarm devices or technical means, such as doors, windows, locks that meet the criteria required in the guidelines. The electromagnetic protection of a system consists in preventing the loss of confidentiality and availability of classified information processed in the data communications devices. The loss of availability can be, in particular, the result of disturbing the operation of ICT devices with high power electromagnetic pulses. The electromagnetic protection of a system or ICT network is ensured, in particular, by placing the ICT devices in controlled access zones, which comply with the requirements for electromagnetic attenuation, and by the use of fibre optic cables, instead of traditional copper cables.

In order to ensure sufficient confidentiality and integrity of classified information being processed, cryptographic protection is used, in a form of encryptors (to encrypt the transmission over networks), firewalls with options to create VPN tunnels or other devices ensuring hardware completion of VPN tunnels.

Ensuring the integrity and high availability of classified information processed in the data communications systems or networks is achieved through redundancy of key system components (servers, databases, application servers), by using backup transmission links, UPS devices and use of power strips preventing from power outages and instability in the power grid. Those elements are crucial for ensuring continuity of the system operation.

The definition of availability is one of the basic categories of security, specifying that authorized persons can use it in the required time and place. Provision of access control to premises and equipment communication system is achieved through technical measures in the form of access control systems, magnetic card readers, entrance gates, and electromagnetic yokes.

22.5 Software Protections

The area requiring special security measures is the electronic environment for electronic data processing in a form of an IT system. Protections used in it are crucial to ensure the security of classified information processed.

Protection and access control to the IT systems is achieved through the use of mechanisms of the system user roles and permissions, the central administration of user accounts in the domain, the use of PKI with certificates issued on the name of registered system users.

The first level of securing the IT systems against unauthorized access is to force a user logs on workstations, using a unique login and password or individual certificate. Since the login and password or certificate is assigned to a particular system user, clear authorization and authentication of the user in the system is ensured. Authentication is the process of verifying the declared identity of a person, device or service involved in the exchange of data.

A user logged in the IT system should have access to only those functionalities and data to which he has been granted permission by the system administrator. User permissions should be checked in multiple stages: from the operating system rights, through the powers given in the application, to the database permissions. Currently, the database servers offer the ability to restrict access to both the individual tables, and single database records. Such a restriction of information visibility causes the system to realize the principle of "the need to know."

Ensuring accountability is one of the basic security functions of ICT systems. In order to meet the requirements, of user activities in the system must leave a trace. Each operation performed by the user must be registered as evidence. It is recommended to store information about who, when and what was done in the system and to what extent. Accountability is achieved by using dedicated systems or functionalities, or built-in database server mechanisms for registration of activities, such as audit logs, database triggers, journal tables, historical tables that store the operations performed, and tables storing the history of objects change in time. The additional (technical) information gathered allow you recreate the object at any point in time, thus providing information about all actions performed by users on the system. Data recorded in the system should be available as evidence, hence it is important to prepare it appropriately in the databases, for the need of presentation and aggregation of reports.

Confidentiality is one of the key safety attributes of classified information that should be especially protected. Appropriate level of confidentiality protection is achieved by using encryption of network connections between the workstations and servers using VPN (SSL connections, access to applications via HTTPS), storing data in the database, on an encrypted partition, or by using encryption of individual database records, protecting access to servers with a user account and password. In the case of exchange of classified information between different security zones, or IT systems, additional security measures should be used in the form of hardware encryptors.

In order to ensure data integrity, mechanisms are used in the form of checksums, correcting codes and certificates of registered users. For the security of classified information, they prevent accidental or intentional distortion of the data during read, write, transmission or storage phases with advanced techniques.

To ensure availability of an IT system and data stored in, care must be taken that the servers work in primary-secondary mode or on a cluster basis, and data be properly processed and stored. A very important element are backups of the database made regularly, and application functions to detect and correct errors, in order to minimize their impact on the continuity of the entire ICT system, processing classified information.

These days, a great threat to the security of information processed in IT systems is the action of mobile and malicious code, and exploiting loopholes in the systems. Malicious code can be entered into the IT system with the aim to disclose data processed in it, especially the classified data. Protection against malicious and mobile code is ensured by using, inter alia, appropriate anti-virus software. Anti-virus protection should be exercised on all components of the system and the ICT network, and it consists in blocking the possibility of executing a code which poses a danger. Protection of systems and networks is achieved by regular updates of anti-virus software (virus definitions), updates to operating systems and database servers with available patches and service packs, blocking unused ports, analysing outbound and inbound network traffic tracking the CVE (Common Vulnerabilities and Exposures) lists and preventing them, running the software with as low system privileges as possible, using firewall software, disabling unnecessary operating system services, or even separation from the external Internet network. An important element is the separation of the functions performed by the database and application servers on separate machines, so that in case of attack on weak points (gaps) of one of the systems/servers, the attackers could not gain control over the entire machine, in which other servers (e.g. database or application servers) can also be installed.

22.5.1 Consequences of Using Powerful Protections in Databases and IT Systems

The Act and its implementing regulations require institutions and individuals to introduce a number of necessary safeguards in their ICT systems, which process classified information. In discussing the tools and methods for protection of classified information, the consequences of their use may not be ignored.

A large number of safeguards and mechanisms applied can affect operation of the entire system and its performance. Multiplying the amount of information processed and stored entail an increased burden on the IT system and the need to find appropriate solutions to this problem. Therefore, a very important element of the project is to prepare the appropriate security of an entire IT system at the design stage.

Planning technical solutions designed to ensure the safety of processed and stored classified information is crucial in estimating the cost of the entire solution. If the system performance requirements are not well estimated, as early as at the design stage, operating problems may be encountered at a later time. The result of this can be both, slowing down the system and its complete blockage.

A large number of security measures also makes the administrations of its elements difficult. Many of the mechanisms introduced to ICT systems and their interaction in the environment is highly interconnected. Changing one of the parameters of the system can affect others, so it is important to prepare appropriate administrative and operating procedures for the entire system, so that, in the event of an emergency, proper system operation can quickly be restored. A large number of configuration elements, a large quantity of logs collected from the heterogeneous environment makes it difficult to manage it, and it affects the number of people needed to administer the entire system. Such additional information, however, (e.g. server or network device logs) are invaluable when searching for errors, analysing abnormal or suspicious behaviour of the system and data communication networks.

The question hast to be therefore asked whether all of those protections must be used in systems that process classified information. They cause complications in designing a solution and significantly affect the cost of the entire solution.

To answer this question, one should realize how important the presence, in such a class of information systems, seems to be for the critical infrastructure or key equipment for proper operation of the system. In the event of an emergency (i.e. an emergency state, flood, fire) the lack of a reserve processing centre, the lack of redundancy of key components might affect the continuity of the whole system operation; the lack of backups can cause permanent loss of important data constituting state or business secrets, which may jeopardize the safety of the organization or the entire state. Therefore, while designing an IT system used to process classified information, the provisions of the Act, implementing regulations and recommendations of DBTI ABW must be strictly complied with. First, they are vital for the compliance with the requirements of the Polish law, secondly, they ensure correct operation of the system and maintaining appropriate levels of security of classified information processed and stored in different situations.

Public Key Infrastructure As The Tool Ensuring IT&T Safety

As mentioned earlier electronic environment for information processing requires application of particular programmary measures supporting protection against unauthorised access, maintenance of information authenticity and integrity as well as accountability of activities. These functions may be successfully fulfilled by appropriately designed, implemented and exploited public key infrastructure, especially electronic signature service operating within its framework.

Electronic signature is based on the concept of Public Key Infrastructure (PKI). It is based on application of the pair of encryption keys as well as digital certificate related to it. For good understanding of PKI operation it is worth quoting a few definitions of basic notions from this field:

- Private key – the first of the pair of encryption keys. It is dedicated to signing data. The key should be available only to one holder (private) hence one should take care of its confidentiality in a particular manner.
- Public key – the second of the pair of keys, related to the private key. It is dedicated to verification of electronic signature. This key should be available to everyone who may potentially use it to verify electronic signature (public).
- Digital certificate – electronic document which attaches public key to a particular person (or institution or another entity). It constitutes the source for information of who is the holder of a particular key. The certificate is electronically signed by the institution which has issued it – so called certifying office. This ensures its reliability (nobody can impersonate the legitimate holder of the pair of keys).
- The period of certificate validity – because of safety reasons the certificates are issued for limited period of time. In case of certificates dedicated to electronic signature it is usually 1-2 years. After expiry date the certificate may be renewed for another period.
- Certification Authority, CA is an entrusted institution which deals with issuing digital certificates. If we trust the authority in the same way we trust all certificates issued by this authority.
- Hierarchy of certification authorities– certification authorities may be hierarchically oriented. This means that the authority on the top of such a hierarchy does not issue digital certificates for users but it only certifies credibility of other authorities directly under it. Subordinate authorities may issue certificates or certify credibility of authorities one step lower in hierarchy. Such a structure is used e.g. in case of qualified signature, broader description on this issue may be found further in hereby document
- Smart card constitutes the most popular and one of the safest ways for ensuring confidentiality of private key. Its construction and software enable reading of private key – the key never leaves the card. In order to produce electronic signature one should send to the card the data dedicated to signing and the returned result is the basis for construction of electronic signature.

In order to use electronic signature one should submit to the certification authority the application for issuing the certificate. The certification authority shall generate at random a pair of encryption keys on the smart card and it shall provide signed by itself digital certificate of public key (also recorded on the card). The certificate shall be valid for example for two years, this means that one may use the card e.g. to produce electronic signature. After the expiry date one should contact the certification authority in order to renew the validity of the certificate for the following years.

Fig. 22.1 Multifunctional smart card

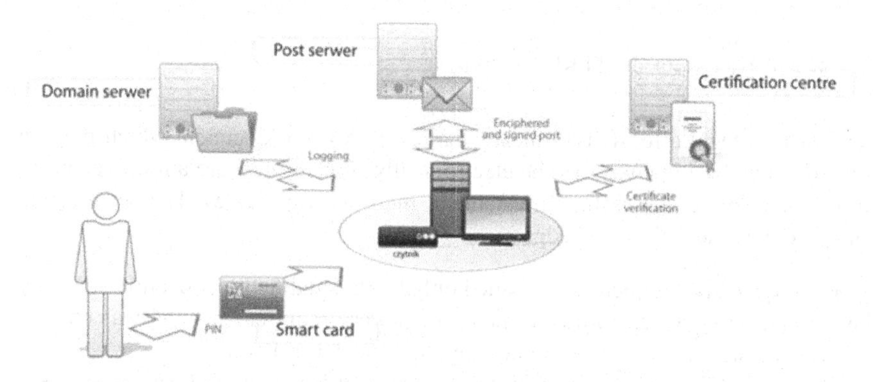

Fig. 22.2 Use of PKI in user's environment

The certificate with the pair of keys may successfully be used for safe authorisation in IT systems as well as signing and encrypting documents. It executes defined at the beginning of this chapter postulates related to the use of PKI as the tool for safeguarding IT infrastructure.

22.6 Electronic Signature and Its Legal Aspect

In order to safely use PKI to safeguard IT&T infrastructure one should use specific norms and standards ensuring high level of reliability and resistance to assault attempts. Application of such norms should be enforced by appropriate provisions of law.

In Polish law there is defined the notion of 'safe electronic signature verified by a qualified certificate'. Coloquially it is called a qualified signature. From technical point of view there is no difference in reference to a standard electronic signature (also referred to as non-qualified signature). The difference is in imposing a range of organisational requirements as well as some technical standards whose goal theoretically is ensuring full safety of such a signature. The binding act on electronic signature states that qualified signature is equal to hand written signature in

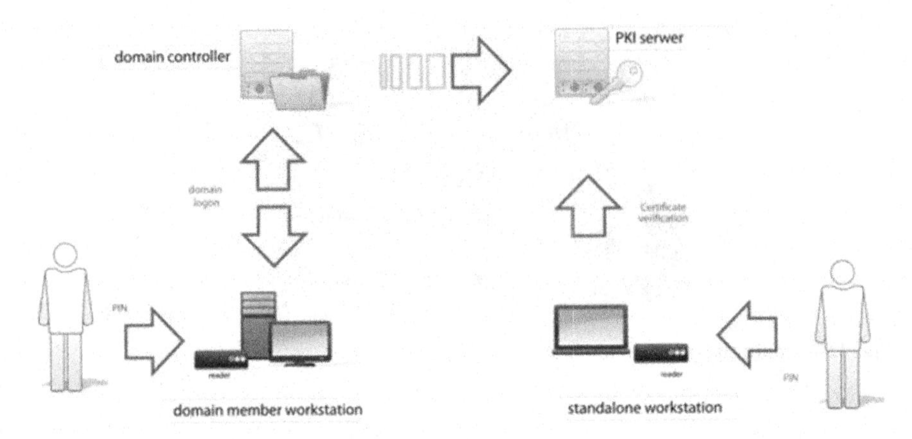

Fig. 22.3 Popular methods of PKI authorisation

reference to legal effects. This means that every kind of docunent which may be signed in handwriting may also be electronically signed (there are a few exceptions from this rule e.g. last will and testament, marriage certificate). The act imposes among others the following hedges:

- qualified certificates are issued only by the qualified certification authority i.e. en enterprise which fulfills a range of statutary requirements and was entered into the register of qualified exhibitors by the Minister of Economy
- qualified certificates are issued only to natural persons (a firm or institution may not apply for such a certificate but must delegate its employees to do this)
- certificates with the pair of keys must be issued on so called 'safe device for producing electronic signature'. Such a device consists of technical component (smart card) as well as software. Such a device must comply with range of norms and standards and its manufacturer must certify that it has ensured such compliance (this is not a subject to verification)

In Poland the function of the main certification authority is fulfilled by the National Certification Centre (NCC). It is a subsidiary of National Bank of Poland, which was entrusted conducting the register of qualified issuers of certificates by the Minister of Economy by virtue of an act. NCC is, therefore, on the top of hierarchy of certification authorities. Below there are particular issuers of qualified certificates and NCC certifies their credibility.

Currently in Poland there act five issuers of qualified certificates:

- Sigillum – Polish Security Printing Works
- KIR - Krajowa Izba Rozliczeniowa
- CERTUM – Unizeto Technologies
- Mobicert
- Safe Technologies

Fig. 22.4 An example of PKI hierarchy

22.7 Popularisation of PKI Technology in Poland

In order to successfully implement IT&T securities basing on public key infrastructure one should take care about popularisation of this technology. The system is safe only when all its users consequently apply the same securing tools. Unfortunately electronic signature has so far not gained popularity.

The history of electronic signature in Poland is almost ten year history as the act on signature came into force at the end of 2001. Despite such a long time popularity of electronic signature, especially among citizens, is very slight. Even in enterprises and public administration in which application was enforced by legal regulations it is used in very little extent and usually as necessary evil. According to authors' intention is was supposed to simplify dealing with everyday matters by the citizens and facilitate communication in administration and business. What causes constitute the basis for such a failure?

Electronic signature is too expensive. The set for production of electronic signature consisting of a smart card with certificate issued for two years as well as card reader is the cost of about PLN 300 – 400. On the other hand the number of files which may be dealt with without leaving home with the use of electronic signature is slight. Until not long ago the only relatively popular services were submitting declarations to Polish Social Insurance Institution (ZUS) and Fiscal Office. In order to enable large group of citizens to submit PIT declarations the Fiscal Office withdraw from the requirement to apply qualified signature. In this context spending a few hundred zlotys on the card with a qualified certificate, which should enable dealing no more than a few files is totally economically unjustifiable. It is impossible to wonder why the service is a niche.

Technical and organisational requirements provided by the Act on signature are very rigorous. According to authors' intention it was supposed to ensure full safety

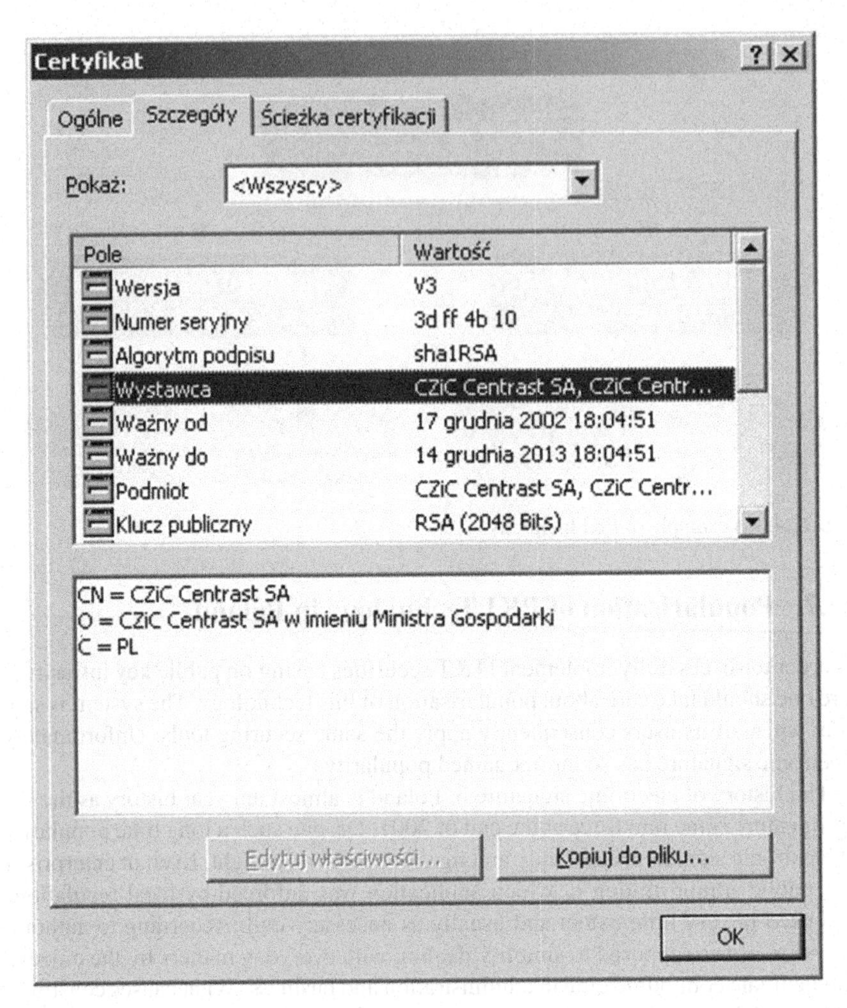

Fig. 22.5 Preview of certificate attributes in Windows system

of electronic signature. Practically, it has caused that fulfilment of these requirements is very costly and it inhibits development of software and hardware dedicated to service of signature. For example complicated procedures of certification of devices (especially smart cards) last many months and they require from producers bearing substantial costs.

Such high requirements in safety are completely unjustifiable. If we compare electronic signature with traditional signature it will turn out that any implementation of PKI (even not fulfilling the highest standards) shall be many times less susceptible to abuses than traditional writing initials with a pen or verification on the basis of template in national ID card.

An obstacle may also be very narrow field of applications of electronic signature. Qualified certificates may currently be issued only for natural persons. Moreover, the act imposes the requirement that producing electronic signature should every time be confirmed with entering PIN. Hence, e.g. mass issuing of invoices or bills in electronic form with signature is impossible. Each such an invoice would have to be separately signed by a specific worker.

Qualified electronic signature may be produced only with the use of special software usually delivered by the issuer of signature. Although many popular programmes render available technical possibility of producing a signature (e.g. Office package, mail programmes, Internet browsers etc. ...) they may not be used for producing a qualified signature.

The act on electronic signature as well as ordinances accompanying it allow for many incompatible formats of electronic signature. Verification of signature produced with the use of software of one of the producers may cause difficulties in the system of another company. It is in conflict with the idea of universality of electronic signature.

22.8 The Possibilities of Development of the Concept of PKI Application

Despite currently dominating unfavourable legal circumstances popularisation of PKI also for protection of IT&T systems is achievable and worth aiming at. It is possible to achieve by means of promoting organisational and legal solutions which shall positively influence popularity of this technology as well as shall enwiden the field of applications including among others wide range of IT&T systems processing confidential information.

Qualified electronic signature has legal capacity on the territory of the whole Poland irrespectively on which document it has been produced. In many cases such a scope and universality of signature is not necessary. Maybe it is worth allowing institutions and enterprises to build its own PKI infrastructure, within which they would issue certificates for their clients or citizens.

Two leading examples are territorial self-government and banking. Many boroughs present very modern approach towards citizen service trying to facilitate it mainly by means of computerisation. Providing the possibility to submit applications and receiving administrative decisions in electronic form makes sense only in case that the majority of citizens use electronic signature. Such boroughs would eagerly issue digital certificates for citizens, even free of charge. The benefit would be mutual, a citizen obtains comfortable method of communication with an authority without bearing additional costs and the authority as such facilitates work getting rid of queuses and limiting the number of processed paper documents.

In similar situation are banks which are more and more frequently communicating with their clients only on the internet. Enabling them to issue certificates dedicated to electronic signature (basing on a similar rule as issuing debit cards) would contribute to development of electronic banking and improvement of safety of transactions. Introduction of such solutions shall enable service of electronic signature in two models:

- sales and implementation of solutions enabling creation and management of PKI infrastructure by a client
- rendering the service constituting management of PKI infrastructure as well as issuing digital certificates for the benefit of the client

Positive influence on popularisation of electronic signature could have softening of some legal and organisational requirements. It is worth proposing more liberal requirements concerning especially devices for producing and verifying electronic signature (understood as hardware with software). Lowering of requirements shall contribute to decrease in prices of devices and thanks to this easier acces to them.

More and more frequently there is raised an issue of abolishing the obligation to apply a technical component such as a smart card for keeping private key related to the certificate.

The act assumes that a holder of qualified certificate keeps his/her private key connected with certificate on a carrier such as smart card. Introduction of the possibility to store the key in central repository of issuer of certificates shall enable execution of electronic signature like e.g. biometric signature:

- pair of encryption keys is generated on the server of a company rendering the service of biometric signature
- the keys are stored in the manner ensuring their safety (accessibility, confidentiality and integrity)
- electronic signature is rendered in the form of service, in which the key holder obtains the access to the function of signature after authorisation with the use of biometric technique (e.g. scan of fingerprints or fingerveins)
- a user obtains access to electronic signature service by means of special terminals accessible in a public place (e.g. city office, bank branch, customer service point, etc. ...) or on the internet with the help of especially designed network application (this requires possessing a biometric reader)

Another important step in development and popularisation of PKI is uniformisation of formats of electronic signature.

It is worth suggesting precision of standards and formats connected with implementation of electronic signature in the form of legal acts for the purpose of ensuring coherence and compatibility of particular systems. The aim is achievement of a situation in which electronic signature may be easily verified in any IT system complying with requirements of the act, irrespectively what tools have been used to produce it. It seems to be justified to allow only one format (probably XaDES) and to deliver a detailed instruction concerning principles of its processing. This will allow to avoid ambiguities connected with different interpretations as well as the way of implementation of the standard by various performers. Experience shows that the key to popularisation of a given standard or protocol is accessibility of tools enabling its processing on various hardware and software platforms (e.g. HTML, PDF, JPEG formats etc.).

Summing up thoughts on development of PKI and electronic signature it is worth noting the following directions:

- enabling enterprises and institutions issuing certificates dedicated to submission and verification of electronic signature within managed by itself PKI infrastructure
- enabling enterprises and institutions ordering a service constituting service of PKI infrastructure and issuing certificates dedicated to producing and verifying electronic signature
- considering such an electronic signature as equal to hand written in matters dealt with by this entrepreneur and his/her client (or office and citizen)
- enabling execution of electronic signature by way of a remote service, not requiring possession of a safe device for producing a signature by a subscriber (certificate holder)

References

1. Act from January 22, 1999 cocerning protection of confidential information (1999)
2. Ordinance of the Chairman of the Council of Ministers dated August 25, 2005 concerning basic requirements of IT&T safety (2005)
3. Recommendations of DBTI ABW concerning software used in systems and networks processing no confidential
4. Act from September 18, 2001anout electric signature (2001)
5. Ordinance of the minister of economy from August 9, 2002 as to specifying detailed mode of creation and issuing cerrification connected with electronic driver (2002)
6. Ordinance of the minister of economy dated August 9 as to specification of detailed mode of creation and issuing certificates connected with electronic signature

Part IV
Interdisciplinary Problems

Chapter 23
Modern MEMS Acceleration Sensors in Tele-Monitoring Systems for Movement Parameters and Human Fall Remote Detection

Pawel Kostka and Ewaryst Tkacz

Abstract. Presented system combines the rapid technological advances in wireless communication and on the other hand progress in sophisticated, miniaturized sensors, to create remote detector of human fall and walking parameters. New MEMs acceleration sensor, with A/D converter, filter and additional peripherals on the miniaturized board was connected to microcontroller system by SPI bus, control system with embedded human fall detection algorithm sends reports and alarms using wireless GSM or GPRS protocol. To obtain detection rules a series of tests were carried out including simulated human fall, controlled lying down, sitting, walking.

23.1 Introduction

Combining of rapid technological advances in wireless communication and on the other hand in sophisticated, miniaturized sensors, gives the opportunities to design system for remote control of human vital function or physical parameters e.g. connected with movement and position. In recent years, technological advances in microelectromechanical-system (MEMS) acceleration sensors have made it possible to design fall detectors based on a 3-axis integrated MEMS (iMEMSő) accelerometer. The technique is based on the principle of detecting changes in motion and body position of an individual, wearing a sensor, by tracking acceleration changes in three orthogonal directions. The data is continuously analyzed algorithmically to determine whether the individual's body is falling or not. If an individual falls, the device can employ GPS and a wireless transmitter to determine the location and issue an alert in order to get assistance. The core element of fall detection is an effective, reliable detection principle and algorithm to judge the existence of an emergency fall situation Modern construction of MEMS accelerometer sensors allow for static

Pawel Kostka · Ewaryst Tkacz
Business Academy in Dabrowa Gornicza, Department of Computer Science
Cieplaka Street 1c, Dabrowa Górnicza, Poland
e-mail: pkostka@polsl.pl, etkacz@polsl.pl

A. Kapczynski et al. (Eds.): Internet - Technical Develop. & Appli. 2, AISC 118, pp. 271–277.
springerlink.com

and dynamic acceleration measurement, thanks to which, they can be used also for other phenomena detection like vibration or inclination. Chosen application fields of modern MEMS acceleration sensors (many in vehicles systems):

- smart air bags
- correction of fuel level indicator while driving over elevation [1]
- automatic positioning of car lights [2]
- alarm systems
- position detection in PDAs, mobile console, smart phones [2], [3]
- hard disc damage prevention while mobile computers falling

Used iMEMS semiconductor technology (by Analog Device) combines micromechanical structures and electrical circuits on a single silicon chip. Using this technology, iMEMS accelerometers sense acceleration on one, two, or even three axes, and provide analog or digital outputs.

23.2 Methods

23.2.1 MEMS Acceleration Sensors Principle of Operation

MEMS accelerometers are produced in the process of surface micromechanics (fig.23.1). Sensors used in presented system detects acceleration by means of inertia of the mass. It corresponds to the simple model of the mass hung on the spring (fig.23.2). In electronic type sensor there is a need to convert the mass movement into electrical signal. A leading world companies like e.g. Analog Devices or Freescale Semiconductor use an acceleration into capacity conversion [5]. Sensors of these manufactures have a special differential capacitor, which capacity changes according the acceleration. It has comb-like structure, where positive and negative capacitor covers are placed alternately (fig.23.3).

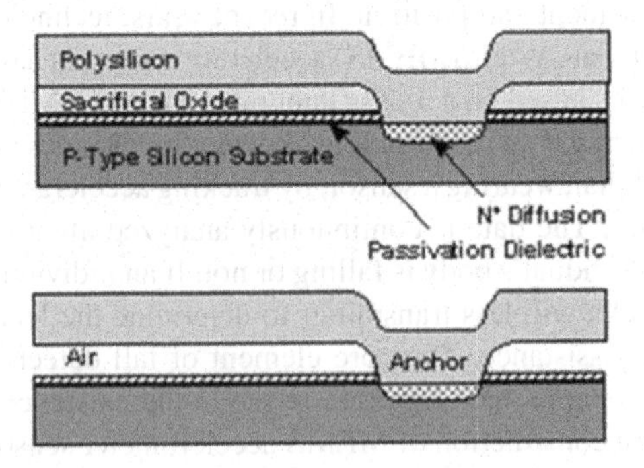

Fig. 23.1 MEMS sensor based on inertial mass effect internal structure [4]

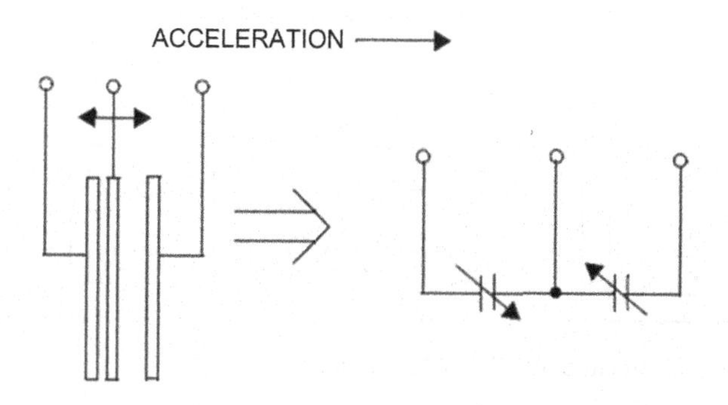

Fig. 23.2 Differential capacitor model [5]

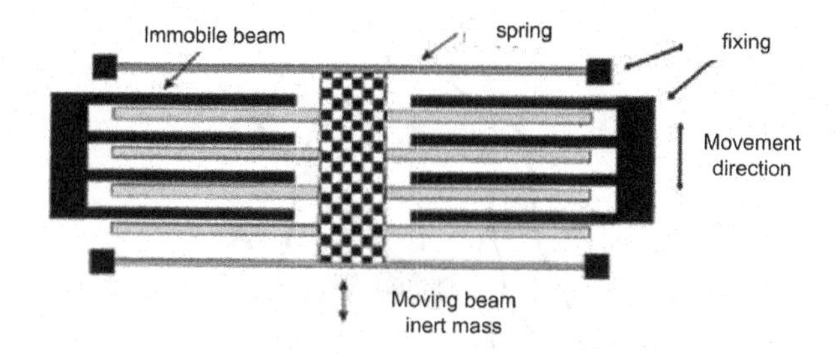

Fig. 23.3 MEMS accelerometer sensor based on inertial mass effect

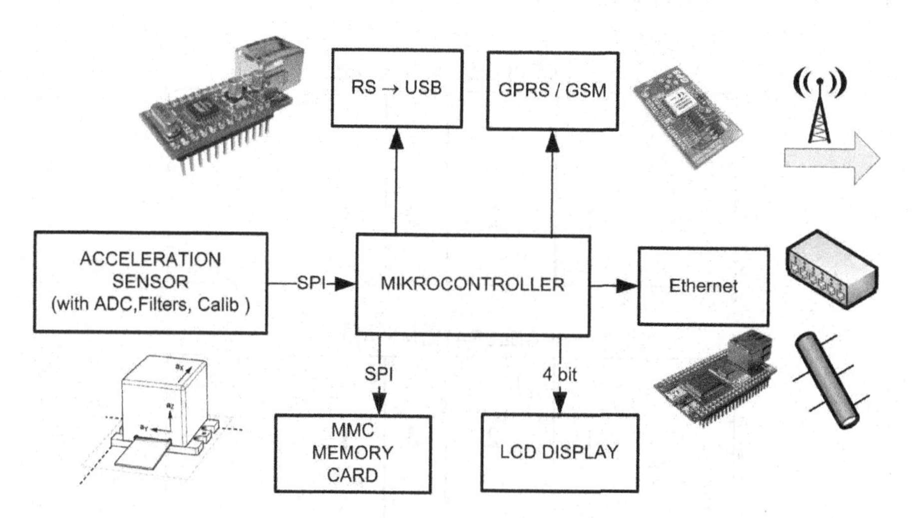

Fig. 23.4 Structure of mobile, data recording system with tele-transmission units

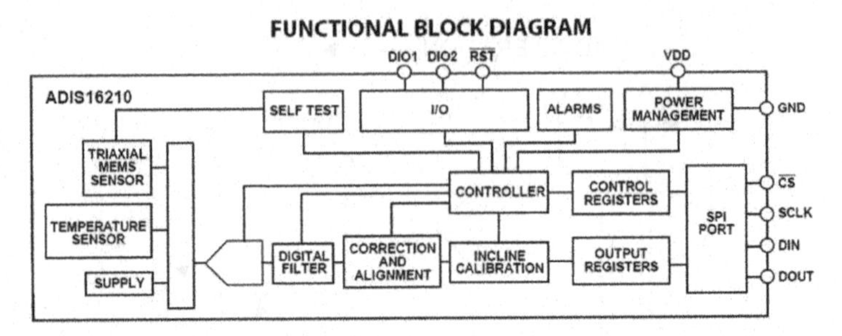

Fig. 23.5 Internal structure of ADIS16210 sensor.

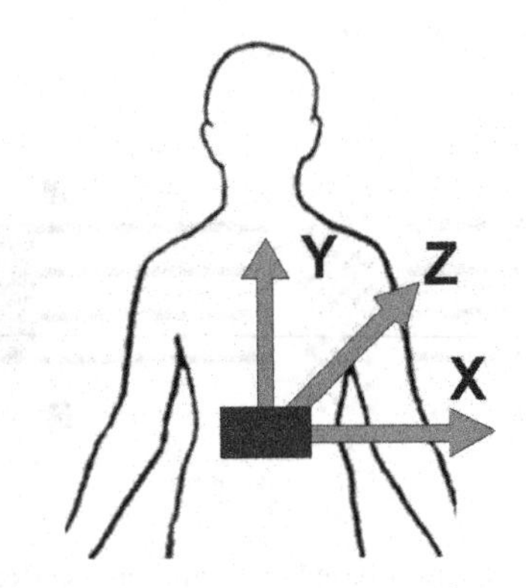

Fig. 23.6 MEMS sensor placement

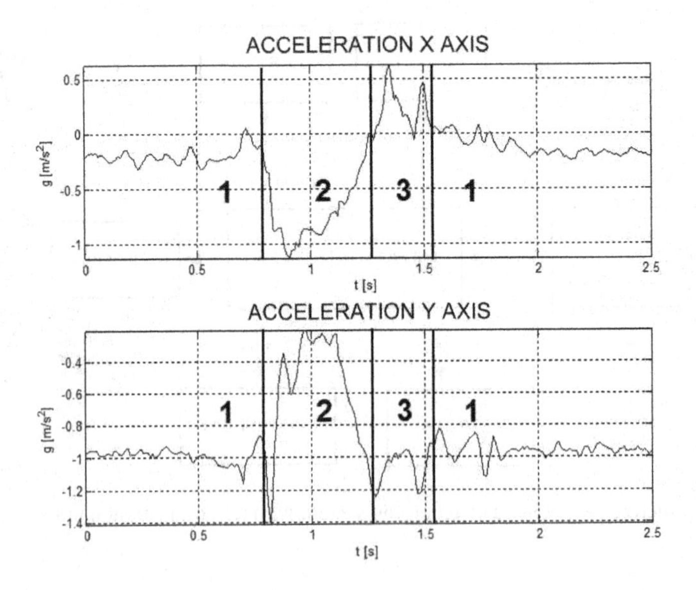

Fig. 23.7 Result: Acceleration while normal walking.

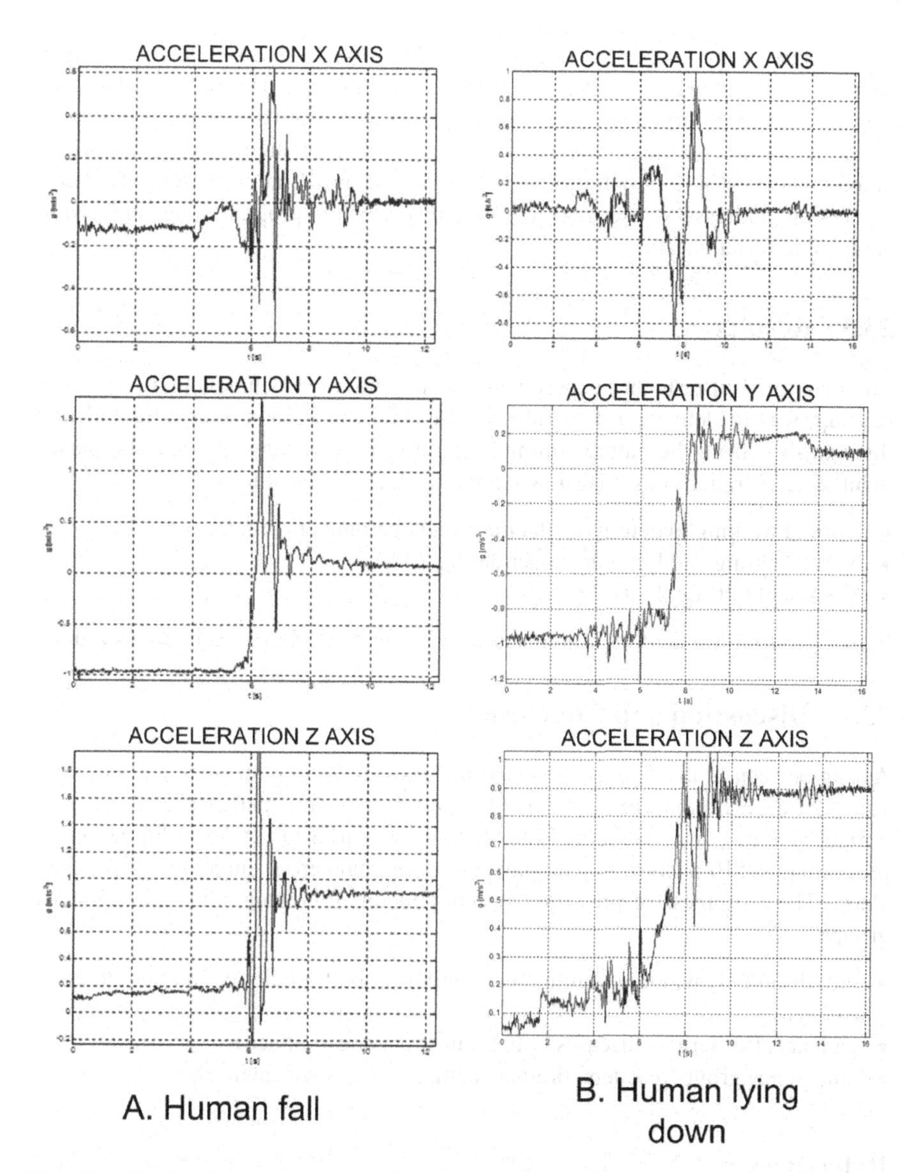

Fig. 23.8 Result: Comparison of acceleration signal while human normal lying down and falling

23.2.2 Mobile Data Acquisition System with Tele-Monitoring Modules

Presented system is an autonomous, battery powered, mobile device for MEMs acceleration signal recording in 3 axis with miniaturized modules supporting Ethernet, GSM and GPRS communication.

The state of the art MEMS accelerator: ADIS-16210 (Analog Device), chosen for this application is 3-axis, high level miniaturized sensor. The ADIS16210 iSensor, which internal structure is presented in fig.23.5 is a digital inclinometer system that provides precise measurements for both pitch and roll angles over a full orientation range of ś180ř. It combines a MEMS tri-axial acceleration sensor with signal processing, addressable user registers for data collection/programming, and a SPI-compatible serial interface.

23.3 Results

The main goal of tests and experiments carried out after technical and functional start-up, was to define a set of rules for detection of human dangerous fall and distinguish it from the natural sitting, lying down or bending actions. The placement of acceleration sensor is shown in the fig.23.6. Types of tests:

- Normal walking to determine the movement parameters (fig.23.7)
- Normal sitting and lying down action (fig.23.8)
- Simulated fall (fig.23.8)

In every test the value of measured acceleration on X, Y, Z axes were compared.

23.4 Discussion and Conclusion

As can be seen in the figures presented in Result section, both slope and absolute values of acceleration during recorded experiments significantly differ in case on normal activities and dangerous fall. It is the material to set up a digital signal processing (DSP) analyse to automatically detect dangerous situation and send an alarm. The usefulness of presented system contains following potential recipients groups:

- old, disable demanding continuous monitoring both in clinical and home environment
- patients after serious diseases (e.g. brain stroke, heart attack)
- emergency group, rescuers, fireman, acting in unknown environment

References

1. Sulouff, B.: Design Design and Development of and Development of Accelerometers and Accelerometers and Gyros,
`http://www.njnano.org/pasi/event/talks/sulouff2.pdf`
2. Liao, W., Zhao, Y.: Using Dual-Axis Accelerometers to Protect Hard Disk Drives,
`http://www.analog.com/library/analogdialogue/archives/39-11/hdd.html`

3. Lemaire, C., Sulouff, B.: Surface Micromachined Sensors For Vehicle Navigation Systems
4. Sulouff, B.: Design and Devellopmentt of Accelerrometers and Gyrros, Tutorial 2B, http://www.njnano.org/pasi/event/talks/sulouff2.pdf
5. Freescale Semiconductor. MMA7260Q. ś1.5g - 6g Three Axis Low-g Micromachined Accelerometer, http://www.freescale.com/files/sensors/doc/datasheet/MMA7260Q.pdf

Chapter 24
Tele-Manipulation System for Minimal Invasive Surgery Support. Prototype for Long Distance Operation

Pawel Kostka and Zbigniew Nawrat

Abstract. System for remote manipulating of endoscopic vision channel as support for minimal invasive surgery is presented. Original control system for telemanipulator working in Master-Slave configuration is implemented on Real Time system and partially as a interface for Master unit on reprogramovable FPGA, Xilinx, where data acquisition and processing algorithm work directly on silicon. An experimental set-up for long distance operation is also described, where surgeon , operator was steering the Master console in Institute of Heart Prostheses in Zabrze and Slave robot arm, equipped with vision channel or coagulation knife carried out operation on pig heart in Center of Experimental Medicine of Silesian Medical University in Katowice (about 30km distance). Results as delays in audio/video channel and motion data transmission were measured. After tele-lab monitoring and control tasks, presented remote manipulation is next step in the program of tele-projects, carried out by Institute of Heart Prostheses Foundation for Cardiac Surgery Development in Zabrze.

24.1 Introduction

Telemanipulator invented for less, minimally invasive cardiac surgery is a computer-controlled device, located between surgeon's hands and the tip of a surgical instrument or endoscopic vision channel. Basic requirements for this device are stable operative field of view, direct surgeon control and high level of precision. Main advantages, motivating the introduction tele-manipulators for minimal invasive surgery support are also:

Pawel Kostka · Zbigniew Nawrat
Institute of Heart Prostheses, Foundation for Cardiac Surgery Development, Zabrze, Poland
Silesian University of Technology, Gliwice, Poland
e-mail: pkostka@polsl.pl, nawrat@frk.pl

A. Kapczynski et al. (Eds.): Internet - Technical Develop. & Appli. 2, AISC 118, pp. 279–286.
springerlink.com

- Scaling of movement between Master console and Slave arm
- Filtering and tremor removing
- Comfort of surgeon work improvement

Since the early 1990s, several surgical robotic systems have been developed [1]. In the field of Minimal Invasive Surgery (MIS), especially two projects should to be mentioned as one as commercial systems : the Zeus system (Computer Motion Inc. [1]) and the da Vinci system (Intuitive Surgical Inc. [1]). The Zeus system is no longer available due to patents regulation.

24.2 Methods

24.2.1 Robin Heart Vision (RHV)-Description of Surgery Manipulator and Mechanical Construction

Mechanical construction of RH Vision (fig.24.1) is based on the prototype surgery telemanipulator Robin Heartő Family, developed and tested in Foundation for Cardiac Surgery Development (FCSD) from 2001. Basic idea of the manipulator RobIn Heart (RH) consists of mechanisms realizing fixed in space "constant point", composed of two closed kinematics' chains [4]. The first loop is in fact a typical parallelogram mechanism, used as a transmission mechanism coupled with the second one realized inverse mechanism. By special connection of two rotations coupled by constant angle internal link, the mechanism can change external angle to approximately 150 degrees. The first degree of freedom (DOF) is driven by electric, brushless motor integrated with planetary gear. The second (range up to 120 degrees doubled system of parallel mechanisms) and third DOFs (the parallel mechanisms eliminates the necessity of using a linear slideway) are driven by brushless motors, roller screws and system of strings. All four DOFs uses Maxonő DCBL motors with hall sensors and digital encoders as a control loop position sensor. The construction makes possible fast and not complicated disconnection the drive part of the bunch from the manipulating part.

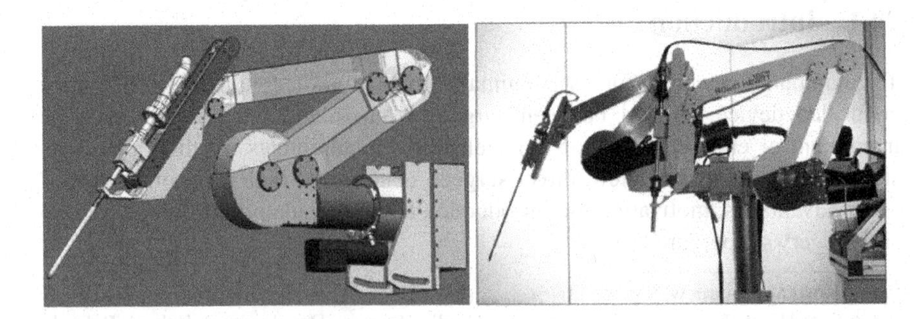

Fig. 24.1 Robin Heart Vision (RHV) Model and Prototype

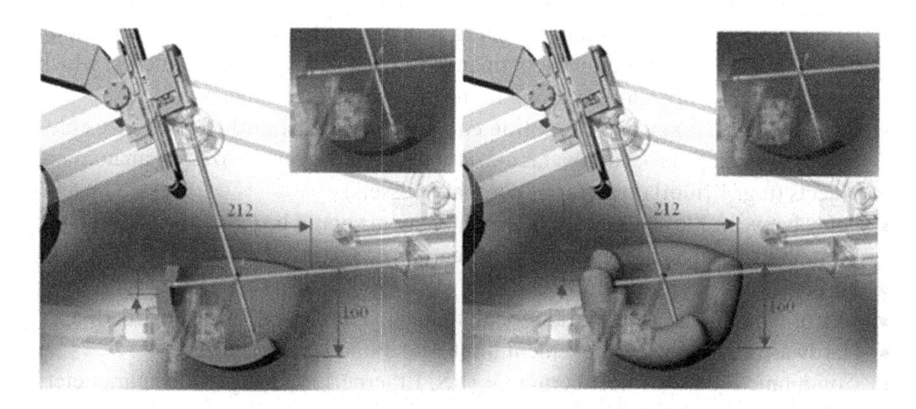

Fig. 24.2 Robin Heart range of operation both for surgery tool and vision chanel

24.2.2 Operation Field and Techniques Analysis for Robotic Supported Cardiac Surgery

To provide all necessary functionality of modern laparoscopic devices, robot Robin Heart gives user a three degrees of freedom to orientate in space, fourth one is responsible for opening and closing jaws of the tool and the fifth one increases the manipulation skills to avoid obstacles, or like Robin Heart allows to work "backwards". Standard laparoscopic device has got a limited mobility and do not offer very sophisticated types of movement that are provided by a robotic systems. To see the differences in a mobility between two various Robin Heart instruments, a tool workspace was calculated for a robot equipped with a standard laparoscopic device and a more advanced robotic instrument. Having a workspace sphere calculated for all of the robot instruments it is very easy to verify the goal of using a suitable device for a proper surgery treatment. Combining this workspace with a geometric position surface we were able to calculate the total range of movement for both robotic instruments inside the patient body (fig.24.2).

24.2.3 Control System for Surgery Telemanipulator: Local Tele-Manipulation

The main idea of control system is common for all described cardio-surgical systems (including USA products: Zeus and daVinci systems). The main task of Master-Slave teleoperator is reliable mapping of surgeon hand movements (temporal values of position/velocity/acceleration or other physical quantity) onto the movements of tool arm, through calculation of control signals for its motors [5]. Technical requirements of RobIn Heart surgical tele-manipulator control system could be listed below:

- Frequency of main program refreshing group signals in the main control loop for translating the Master arm commands into the Slave arm movements, which ensures fluent work should belong to the 200-1000 [Hz] range.
- Positioning accuracy and resolution equals at least 0.1 [mm] to guarantees precision of surgery procedures, taking into account the small sizes of anatomical objects (e.g. 1 [mm] diameter of coronary vessels)
- delay between Master and Slave arm movement should be lower then acceptable limit 330[ms] to fulfil eye-hand coordination
- Surgeon hands tremor elimination
- Mirror movements (like in traditional, manual laparoscopy) effects reduction
- Hardware and software movement limit detection on particular axis
- Communication with host computer (RS, Ethernet) to change work parameters and monitor current state of the system
- Optionally, introduction of force feedback with the possibility of scaling of the force (or others: audio-visual, termical or mechanical - vibrations) sense, passing to operator

Original control system for telemanipulator working in Master-Slave configuration is implemented on Real Time system and partially as a interface for Master unit on reprogramovable FPGA, Xilinx, where data acquisition and processing algorithm work directly on silicon (Hardware from National Instrument with software written in Labview, graphical programming environment(fig.24.4),(fig.24.5).

A low level motor driver units with PID regulators for motors with planear gers, hall sensors and digital encoders (in feedback control loop) (Maxon DCBL motors, EC ECPowerMax family) was developed. A CAN type net of distributed control units was prepared, where every motor unit assigned to particular DOF has its control PID unit with very advanced communication and safety systems (EPOSő , Maxon) placed next to it.

Designed structure of control system is redundant to improve the safety conditions, where apart from basic information channel the second, alternative one added. Pararelly two control channels compute the same required data but using different / doubled sensors (both of Master and Slave tool). In every iteration cycle two groups

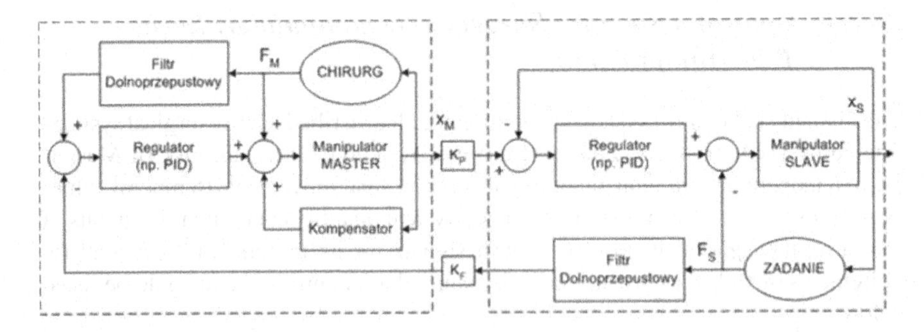

Fig. 24.3 Time of delay measurement both for audio/video and manipulation data stream

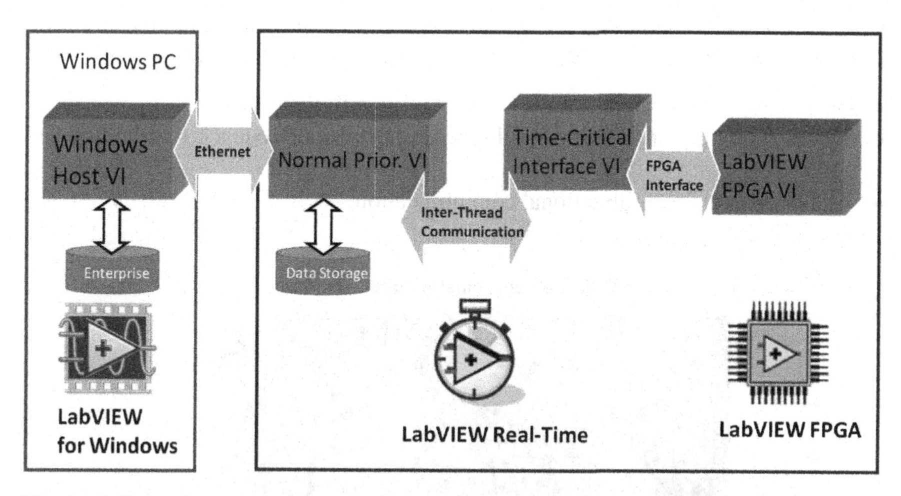

Fig. 24.4 General structure of control system implementation on Industrial type system with Real Time Operation System (RT-OS). Three components: RT Unit, FPGA card, PC application as a terminal

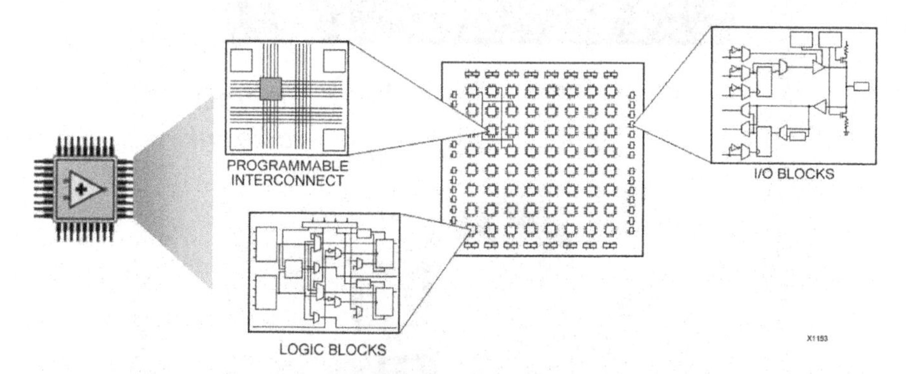

Fig. 24.5 Structure of reprogramowavable FGPA module, which allows to build "system on chip" for specified application

of computed data are compared and in case of difference greater then set limit, immediate stop command with the highest proritet is send to all motors to detect and prevent possible emergencies.

24.2.4 Long Distance Tele-Manipulation Experiment

Control system,prepared for long distance tele-manipulation experiment was created by network-shared project, where Master and Slave part was divided and worked separately using special network-shared variables as communication objects. Several main data transmission channel can be listed (fig.24.6):

- Data manipulation stream (from Master console To Slave arm)
- Video channel, which guarantee the necessary visual feedback (from Slave To Master)
- Audio channel for bi-directional communication.

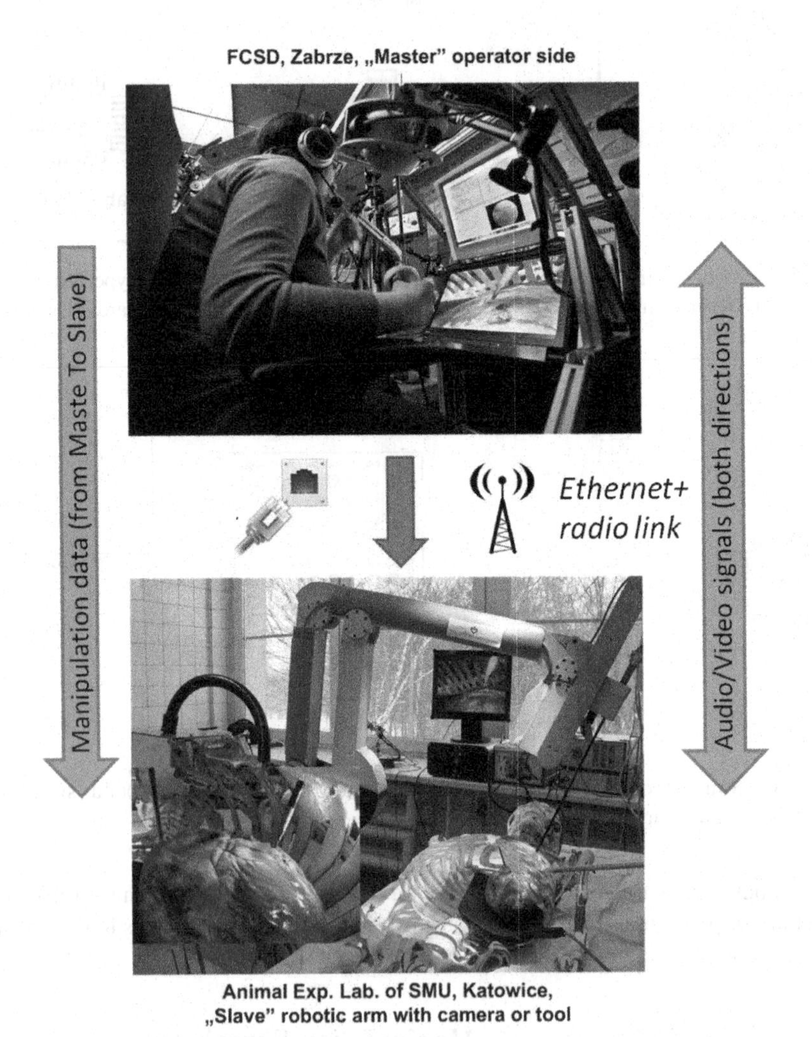

Fig. 24.6 Steering console equipped with Haptic device with Surgeon Operator (Zabrze) and Robin Heart arm with coagulation knife (Katowice)

Fig. 24.7 View on the distance during tele-operation.

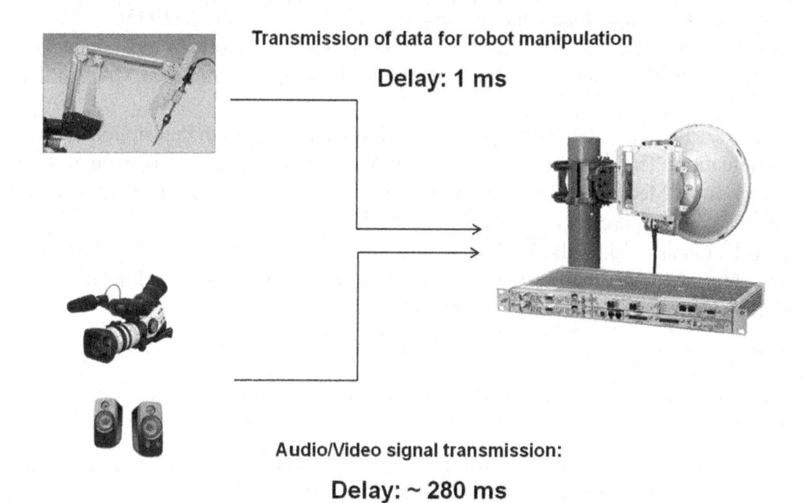

Fig. 24.8 Time of delay measurement both for audio/video and manipulation data stream

24.3 Results

An experimental set-up for long distance operation was developed and successfully tested, where Surgeon - Operator (dr Joanna Sliwka-Los, Surgeon from Silesian Center of Heart Diseases) was steering the Master console in Institute of Heart Prostheses in Zabrze and Slave robot arm, equipped with vision channel or coagulation knife carried out operation on pig heart in Center of Experimental Medicine of Silesian Medical University in Katowice (about 30km distance) (fig.24.7), (fig.24.6).

Results assumed as delays in audio/video channel and motion data transmission were measured (fig.24.8).

24.4 Discussion and Conclusion

After tele-lab monitoring and control tasks described in previous papers, presented here remote manipulation is next step in the program of tele-projects, carried out by Institute of Heart Prostheses Foundation for Cardiac Surgery Development in Zabrze. Apart presented medical application (MIS surgery support), from technical point of view, developed system can be used also in other field, including industrial tele-manipulating in hard accessible places. While manipulation data with delay not increasing 1[ms] is very good result, the video data transmission reached the delay value on the edge of possible limit (to guarantee the good enough coordination).

References

1. Taylor, R., Stoianovici, D.: Medical robotics in computer-integrated surgery. IEEE Trans. Robot. Automat. 19(5), 765–781 (2003)
2. Sackier, J., Wang, Y.: Robotically assisted laparoscopic surgery: From concept to development. In: Computer-Integrated Surgery, pp. 577–580. MIT Press (1995)
3. Guthart, G., Salisbury, J.: The intuitive telesurgery system: Overview and application. In: Proc. IEEE Int. Conf. Robotics and Automation (ICRA), San Francisco, CA, April 2000, pp. 618–621 (2000)
4. Nawrat, Z., Kostka, P.: Polish Cardio-robot Robin Heart, System description and technical evaluation. The International Journal Of Medical Robotics And Computer Assisted Surgery. Int J. Med. Robotics Comput. Assist. Surg. 2, 36–44, Published online March 6, 2006 in Wiley InterScience (2006), doi:10.1002/rcs.67
5. Mohr, F., Onnasch, J., Falk, V., Walther, T., Diegeler, A., Schneider, F., Autschbach, R.: The evolution of minimally invasive valve surgery–2 year experience. Eur. J. Cardiothorac. Surg. 15(3), 233-8; discussion 238-9 (1999)